T0222274

Engineers for Korea

Synthesis Lectures on Global Engineering

Editor
Gary Downey, *Virginia Tech*
Assistant Editor
Kacey Beddoes, *Purdue*

The Global Engineering Series challenges students, faculty and administrators, and working engineers to cross the borders of countries, and it follows those who do. Engineers and engineering have grown up within countries. The visions engineers have had of themselves, their knowledge, and their service have varied dramatically over time and across territorial spaces. Engineers now follow diasporas of industrial corporations, NGOs, and other transnational employers across the planet. To what extent do engineers carry their countries with them? What are key sites of encounters among engineers and non-engineers across the borders of countries? What is at stake when engineers encounter others who understand their knowledge, objectives, work, and identities differently? What is engineering now for? What are engineers now for?

The Series invites short manuscripts making visible the experiences of engineers and engineering students and faculty across the borders of countries. Possible topics include engineers in and out of countries, physical mobility and travel, virtual mobility and travel, geo-spatial distributions of work, international education, international work environments, transnational identities and identity issues, transnational organizations, research collaborations, global normativities, and encounters among engineers and non-engineers across country borders.

The Series juxtaposes contributions from distinct disciplinary, analytical, and geographical perspectives to encourage readers to look beyond familiar intellectual and geographical boundaries for insight and guidance. Holding paramount the goal of high-quality scholarship, it offers learning resources to engineering students and faculty and working engineers crossing the borders of countries. Its commitment is to help them improve engineering work through critical self-analysis and listening.

Engineers for Korea
Kyonghee Han, Gary Lee Downey
June 2014

Designing Development: Case Study of an International Education and Outreach Program
Aditya Johri, Akshay Sharma
January 2013

Merging Languages and Engineering: Partnering Across the Disciplines
John M. Grandin
January 2013

What is Global Engineering Education For? The Making of International Educators, Part III
Gary Lee Downey, Kacey Beddoes
November 2010

What is Global Engineering Education For? The Making of International Educators, Part I & II
Gary Lee Downey, Kacey Beddoes
November 2010

Engineers for Korea
Kyonghee Han and Gary Lee Downey

ISBN: 978-3-031-01000-2   paperback
ISBN: 978-3-031-02128-2   ebook

DOI: 10.1007/978-3-031-02128-2

A Publication in the Springer series
SYNTHESIS LECTURES ON GLOBAL ENGINEERING # 5
Series Editor: Gary Downey, Virginia Tech
Assistant Editor: Kacey Beddoes, Purdue

Series ISSN 2160-7664 Print 2160-7672 Electronic

Engineers for Korea

Kyonghee Han
Yonsei University
Gary Lee Downey
Virginia Tech

SYNTHESIS LECTURES ON GLOBAL ENGINEERING #5

For those engineers who invite us into their lives

ABSTRACT

"The engineer is bearer of the nation's industrialization," says the tower pictured on the front cover. President Park Chung-hee (1917–1979) was seeking to scale up a unified national identity through industrialization, with engineers as iconic leaders. But Park encountered huge obstacles in what he called the "second economy" of mental nationalism. Technical workers had long been subordinate to classically trained scholar officials. Even as the country became an industrial powerhouse, the makers of engineers never found approaches to techno-national formation—engineering education and training—that Koreans would wholly embrace.

This book follows the fraught attempts of engineers to identify with Korea as a whole. It is for engineers, both Korean and non-Korean, who seek to become better critical analysts of their own expertise, identities, and commitments. It is for non-engineers who encounter or are affected by Korean engineers and engineering, and want to understand and engage them. It is for researchers who serve as critical participants in the making of engineers and puzzle over the contents and effects of techno-national formation.

KEYWORDS

engineers, engineering, Korea, Korean engineers, Korean engineering, engineering education, dominant images, dominant practices, national identity, gender in engineering, women in engineering, techno-national formation, scholar official, Park Chung-hee, industrialization, Korean miracle, economic competitiveness, globalization, critical participation

Contents

Preface and Acknowledgments xiii

1   What Are Korean Engineers For? 1
    What is a Korean Engineer? 7
    Government Struggles to Establish Sovereignty 14
    Koreas and Engineers 18
    References 20

2   Five Koreas Without Korean Engineers: 1876–1960 23
    Late Joseon Disinterest in Physical Labor: 1876–1897 25
    Responding to the Threat from Japan: 1897–1910 26
    Low-Level Technicians for the Japanese Empire: 1910–1945 31
    No Place for Engineers in an Agrarian Vision: 1945–1948 38
    Rebuilding Again Without Engineers: 1948–1960 41
        A Matter of Individual Interest and Ambition 47
    References 49

3   Technical Workers for Light Industry: 1961–1970 53
    A Program in Two Parts 54
    Technical Soldiers for Industrial Development 57
        Initial Attempts to Scale Up Technical Education 59
        Higher-Level Experts for Exports 62
    Resistance in the "Second Economy" 66
    References 73

4   Engineers for Heavy and Chemical Industries: 1970–1979 77
    Making Heavy Industry Korean 79
    Vocational Graduates for Specific Industries 88
    Rational Scientist-Engineers for Leadership 90
    Korean Miracle? Continuing Struggles in the Second Economy 93
    References 97

5   Loss of Privilege and Visibility: 1980–1998 101
    Rationalizing the Economy 102
    Engineers Lose the Spotlight: 1980s 108

Coordinated Creativity?.   113
Competitive Self-Development or an Organized Profession? 1990s.   119
References   126

6   Engineers for a Post-Catch-Up Korea? **131**
Scaling Up an Image of Crisis.   132
Uneven Support from Successive Governments   135
The Continuing Struggles of Women Engineers   138
Military Practice and the Dominant Image of Engineering   141
New Images Scaling Up?.   144
References   145

7   Engineers and Korea **149**
Korean Engineers and the Scholar-Official.   151
Critical Self-Reflection and Critical Participation   157
References   158

Index **161**

Author Biographies **175**

Preface and Acknowledgments

We found we were traveling in the same direction. The occasion was the first workshop of the International Network for Engineering Studies (INES) held in Blacksburg, Virginia in 2006. Kyonghee was conducting a study of the "science and engineering crisis," a hot topic in Korea, as part of her research on the identities of Korean engineers. Gary had long been teaching a module on the emergence of engineers and engineering across Korea in his Engineering Cultures course. Drawing on interviews, secondary sources, and assistance from other scholars, he too focused on the identities of Korean engineers.

Subsequent to that workshop, Gary helped Kyonghee edit her manuscript on the science and engineering crisis for publication in the INES/Routledge journal *Engineering Studies*. Kyonghee invited Gary to contribute an article to the *Engineering Education Magazine* published by the Korean Society for Engineering Education. The issues we discussed during many exchanges led Gary to propose a jointly authored volume for this series, suggesting that it might benefit a broad range of readers interested in engineers and engineering in Korea. Kyonghee readily agreed, happy at the prospect of sharing our insights and questions with others. Looking back now over nearly 200 messages from just the past two years, each of us expresses deep gratitude to the other for a collaboration that has been unceasingly exciting and fulfilling, both professionally and personally.

As the introductory and concluding chapters explain, we want this book to help engineering students and working engineers, both inside and outside of Korea, become better critical analysts of their own identities, expertise, and commitments. Such is the first step in learning to work more effectively with both engineers and non-engineers who define problems differently, as well as learning to assess the effects of engineering work as others see and experience them. We also want to further interdisciplinary research in engineering studies and its environs. In particular, we want to call attention to the range of ways that the makers of engineers for Korea have framed and enacted techno-national formation. For colleagues interested in participating critically in engineering formation and work, this book raises and wrestles with a challenging issue. Might the prospects for efforts to formulate and scale up new images and practices of engineers and engineering regularly, if not always, depend upon taking into account localized images of techno-national formation and practice that engineers have advanced, embraced, and accepted as given?

Each of us owes thanks to people and organizations that supported us prior to and during this study. Together we thank Kim Woo Sik, Lee Jang Gyu, Lee Jongmin, Lee Dong Jin, Yi Youngok, and Yoon Woo Young for sharing with us stories of their careers and lives. Thanks also to Song

Sung Soo, Lee Eun Kyung and Kacey Beddoes for serving as reviewers for an earlier version of this book. Their detailed questions, comments, and suggestions all greatly improved the quality of the manuscript.

This volume is the product of a genuine collaboration, to which each of us contributed equally. One outcome in that spirit was our decision to list Kyonghee's name first.

Some of our acknowledgements must appear in the first person, in order to best convey the emotional force of our appreciation.

Kyonghee: I owe a huge thanks to Youn Dae hee for suggesting I seek employment in an engineering school, and for showing patience and trust as I learned to make my way through it. My teacher, Kim Yong hak, has made sure I did not lose a sense of tension and questioning as a sociologist. I express my gratitude to my colleagues who have helped me enter the world of engineering and engineers, always answering my questions. I won't forget the passion and consideration provided by professors Heo Jun-haeng, Cha Sung Woon, Yun Ilgu, Lee Kang Taek, Kim Taeyeon, Kang Hojeong, Kim Hyung Chul, Kim Moon Kyum, Lee Sang Jo, Lee Jaiyong, Min Dong Jun, Cho Hyung hee and Kang So Yeon of Yonsei University. Special thanks are owed to Choi Moonhee, who has always been willing to review unrefined thoughts and writings, and offer suggestions on what should be revised or further developed.

The Korea Institute of Advanced Technology provided a great deal of financial support for this study. Along with Gary, I thank all the organizations that provided photos and valuable data. Thanks in particular to Go Moon-ok, who made available pictures from his own collection. Thanks as well to my friends Choi Young, Sohn Hee-yeon, and Song Sol, who prayed for and encouraged me whenever I was drained.

Finally, my contributions to this volume were made possible by the patience and love of my family. My heartfelt love and thanks go to my husband Shin Kang Ho and my daughters Hwasoo and Eunsoo. I cannot thank enough my mother Kim Ki Suk, who has always filled my empty space.

Gary: For me, this book takes an important step in a career-long project to participate critically in the making and practices of engineers. I care about engineers because they explicitly link the production and expression of quality knowledge and expertise to normative ideals of service. The linkages are interesting intellectually and important personally. All too often, however, engineering pedagogy and practice are self-limiting. They cloud, complicate, and challenge what engineers frequently presume is simple, straight-forward, and self-evident. I am both engineer and ethnographic listener. My contributions to this book, a historical ethnography, constitute a step in highlighting critical self-analysis as a vital engineering practice.

I greatly appreciate the thoughtful and enthusiastic support for this project I received from colleagues committed to provoking engineers via transnational learning. I thank in particular Stephanie Adams, Atsushi Akera, Michael Alley, Dianne Atkinson, Sigrid Berka, Thomas Carter,

Lester Gerhardt, John Grandin, Tom Litzinger, Phil McKnight (deceased), Joseph Mook, Mark Rectanus, Richard Vaz, and Kent Wayland.

I thank Kuo-Hui Chang, Chris Hays, Gouk Tae Kim, SoYeon Park, Kim Gouk Tae, and Masanori Wada for excellent research assistance that provided valuable data for this project. I also appreciate the many conversations and exchanges about engineers and engineering I have had over the past decade with dozens of scholars in engineering studies, science and technology studies, philosophy of engineering, history of engineering, engineering and social justice, women's and gender studies, and engineering education research.

I would like to acknowledge and thank Kacey Beddoes for an amazing multi-dimensional relationship that for me has been pure pleasure. Thanks especially to her company, Education Research 360, for producing a valuable instrument, including scenario-based evaluations, for assessing student learning from this book.

I have benefitted greatly from many faculty and graduate student colleagues and friends who have helped me improve the undergraduate course Engineering Cultures and drew on it in their own projects. I especially thank Juan Lucena, Brent Jesiek, and Matt Wisnioski. Thanks also to the nearly thirty graduate students who, while helping advance the learning of thousands of students, have advanced my own as well.

My contributions to this book are based upon work supported by the National Science Foundation (NSF) under Grant No. DUE1022898. Thanks go to Doris Shelor for her precise, cheerful, and reliable assistance implementing this grant. As required, I assert that any opinions, findings, and conclusions or recommendations expressed in this material are those of the authors and do not necessarily reflect the views of NSF. I do, however, recommend to NSF and the U.S. Congress that NSF continue to fund social and behavioral research at increasing levels, in parallel with other directorates. Critical self-analysis is an essential feature of all quality scientific work.

Finally, I enjoy my work because my days are filled with the love and support of my family, both nuclear and extended. Loving thanks to the many Downeys, Harrises, and Zeilfelders out there for your lifelong support. Above all, thank you Marta, my partner; Jamie, Megan, Michael and Leah, my children; and Nicholas, Andrew, and Casey, my grandsons, for the privilege of having you and your love in my life.

Figures

Figure 1.1: Korean peninsula, 2013.
Source: Map data ©2013 AutoNavi, Google, SK planet, ZENRIN. 15

Figure 1.2: The Three Kingdoms and Gaya (5th Century). Source: www.korea.net. 16

Figure 2.1: Dasan Jeong Yak-yong and the pulley (*Geojunggi*) he designed to
construct Hwaseong Fortress. Source: Jeollanam-do Gangjin-gun. 24

Figure 2.2: The first telephone exchange in Korea.
Source: The City History Compilation Committee of Seoul. 29

Figure 2.3: Field practicum in the 1930s.
Source: The City History Compilation Committee of Seoul. 34

Figure 2.4: Kyeong-seong Textile.
Source: The City History Compilation Committee of Seoul. 36

Figure 2.5: Dr. Ree Tai-kyue (1902–1992) and Dr. Li Seung-ki (1905–1996).
Source: Yuksa-Bipyeongsa Editorial Board 2008: 176. 41

Figure 2.6: Interior facilities of the Office of Atomic Energy (1960).
Source: National Archives of Korea. 44

Figure 2.8: Ceremony to deliver relief supplies from the United Nations, 1959.
Source: National Archives of Korea. 46

Figure 2.9: Sibal, the first assembled car in Korea, by the Kukjecharyang
Company, 1957. Source: National Archives of Korea. 47

Figure 3.1: General Park Chung-hee (1917–1979).
Source: National Archives of Korea, JoongAng Ilbo Photo Archive. 53

Figure 3.2: President Park Chung-hee (1917–1979).
Source: National Archives of Korea. 56

Figure 3.3: Groundbreaking ceremony for the Ulsan industrial area, February 3, 1962.
Source: National Archives of Korea. 60

Figure 3.4: President Park presenting a certificate of appointment to Choi
Hyung-sub, President of KIST. Source: Korea Institute of Science and Technology. 64

Figure 3.5: Declaration of the National Charter of Education, December 5, 1968.
Source: National Archives of Korea. 69

Figure 3.6: Burning incense at the opening ceremony of Gyeong-bu Expressway.
Source: National Archives of Korea. 71

Figure 4.1: President Park's Banner at Busan Mechanical Engineering High School.
Source: http://gomoonok.com. 78

Figure 4.2: Chung Ju-yung. Source: Hyundai Motor Company. 80

Figure 4.3: The 500 KRW note circulating in 1971.
Source: Bank of Korea, Money Museum. 81

Figure 4.4: 1974 Launching ceremony of Atlantic Baron (260,000 ton oil tanker)
by Hyundai Shipbuilding Company. Source: National Archives of Korea. 81

Figure 4.5: President Rhee Syng-man.
Source: The memorial association for President Rhee Syng-man. 82

Figure 4.6: The 1976 ignition ceremony of POSCO's second blast furnace.
Source: National Archives of Korea. 84

Figure 4.7: Korean workers in the LG radio factory in 1962.
Source: National Archives of Korea. 85

Figure 4.8: Park Tae-jun. Source: POSCO. 87

Figure 4.9: President Park encourages researchers while visiting KIST in 1967.
Source: National Archives of Korea. 92

Figure 5.1: President Chun Doo-hwan. Source: National Archives of Korea. 102

Figure 5.2: Ulsan plant view (left) and assembly line (right) in the 1980s.
Source: Hyundai Motor Company. 104

Figure 5.3: Electronic Telecommunications Research Institute (1995).
Source: ETRI 35-Year History. 106

Figure 5.4: Lucky Goldstar in 1970. Source: National Archives of Korea. 108

Figure 5.5: Democratization movement in July 1987.
Source: Korea Democracy Foundation and Kyunghyang Shinmun. 109

Figure 5.6: Labor strike at Ulsan Industrial District in 1987.
Source: Hyeonjang Silcheon Labor Alliance. 110

Figure 5.7: Official mascot of the Seoul 1988 Olympic Games.
Source: Seoul Olympic Museum. 111

Figure 5.8: President Roh Tae-woo. Source: National Archives of Korea. 112

Figure 5.9: President Kim Young-sam. Source: National Archives of Korea. 114

Figure 5.10: Koo Cha-kyung. Source: LG Group. 117

Figure 5.11: President Kim Dae-jung. Source: National Archives of Korea.   122

Figure 5.12: Proportions of women working and completing education by
age group, 2004. Source: KOSTAT, State Statistics.   124

Figure 6.1: President Roh Moo-hyun received representatives of science and
engineering sectors in *Cheongwadae*. Source: Presidential Archives.   131

Figure 6.2: Kim Woo-sik, Deputy Prime Minister, speaks at NAEK award ceremony.
Source: National Academy of Engineering of Korea.   136

Figure 6.3: Leadership camp at Yonsei University organized by the local Women
in Engineering Program. Source: College of Engineering, Yonsei University.   140

Tables

Table 1.1: Classifications of technical workers by the Economic Planning Board.
Source: Ministry of Science and Technology, *Gwahakgisul Yeongam*
(*Science and Technology Annals*) 9

Table 1.2: Classifications of technical workers by the Ministry of Science and
Technology and 1973 Qualification Act.
Source: Ministry of Science and Technology, *Gwahakgisul Yeongam*
(*Science and Technology Annals*) 11

Table 1.3: Classifications of research workers by the Ministry of Science and
Technology.
Source: Ministry of Science and Technology, *Gwahakgisul Yeongam*
(*Science and Technology Annals*) 12

Table 1.4: Gender composition of engineering graduates, 1980–2005.
Source: Ministry of Science and Technology, *Gyoyuk Tonggye Yeonbo*
(*Statistical Yearbook of Education*) 12

Table 1.5: Numbers of researchers in five countries[28] 14

Table 3.1: Fields of study for Rhee and Park administrators.
Source: Han Seung-jo 1975:7[7] 56

Table 5.1: Number of employees by industry, 1965–1995 120

Table 5.2: Suicide rates in Korea during the 1990s.
Source: www.kostat.go.kr (Statistics Korea) 121

CHAPTER 1

What Are Korean Engineers For?

On April 21, 2002, the 34th anniversary of Science Day in the Republic of Korea, the Korean Federation of Science and Technology Societies (KOFST) issued a call for public action. It challenged the Korean government by releasing a "Declaration of Crisis in Science and Technology." "[F]or years," the declaration observed, "our leaders have claimed that 'science and technology are together the country's life-line.'" "But," it asserted, "they have neglected to create a society in which scientists and engineers are nurtured and recognized. ... [W]e can't stop feeling deceived." "[A]s our hopes have crumbled," it warned, "we can no longer tolerate our present situation in which five million scientists and engineers working silently in their classrooms, labs, and the field are neglected."[1]

The year 2002 witnessed a significant escalation in national discussion and debate over the claimed presence of a "crisis" in science and technology. Diverse groups weighed in with their viewpoints. Some groups claiming allegiance to the liberal arts, for example, argued that the situation was entirely natural and expected. Groups representing engineers and scientists, not surprisingly, found it new and threatening. Most interesting for our purposes were the contrasts that different generations of engineers expressed in their arguments. Self-appointed representatives of elder and younger generations tended to agree about the existence of a crisis, but they disagreed greatly about the causes and possible solutions.

In 2012, the National Academy of Engineering of Korea articulated the continuing concerns of engineering leaders by publishing the views of ten of its members. All took for granted that the remarkable industrial expansion in Korea from the 1960s through the 1980s depended upon engineers and scientists who worked diligently, competitively, and with great passion in service of their country. The key image was "hungry spirit," or hungry *Jeong-shin* 헝그리 정신. Hungry *Jeong-shin* (spirit) had been originally used for athletes, especially boxers, who climbed all the way to the top with an indomitable will, without any support, as they were born to poor families. Hungry *Jeong-shin* then came to refer to people's passion and attitude to achieve success by maximally utilizing their own resources, even if there was no breakthrough in a fierce competition, not being

---

[1] Korea Federation of Science and Technology Societies (KOFST), "Declaration of Crisis in Science and Technology," 2002, 9. As we will see below, it has frequently been appropriate in Korea to speak of engineers and scientists together. Following local practice, we put surnames first for Korean authors and other figures. We also use full names, since so many have the surname Kim or Lee. For English spellings, we follow the rules of the National Institute of the Korean Language (http://www.korean.go.kr/eng/index.jsp). Kyonghee Han translated Korean-language sources into English, and Gary Downey edited the translations into colloquial English.

discouraged.[2] Yun Jong-yong, former CEO of Samsung Electronics and then-Chairman of the Presidential Council on Intellectual Property, used the image to highlight the social contributions of Korean engineers:

> Engineers were crucial for the modernization of Korea. In school, we probably learned one-tenth of what current students learn. What we had was a hungry spirit and a sense of mission. Those who are now in their sixties and seventies had contributed considerably to industrialization. How could it have been possible without engineers? [3]

Kwon Oh-kyung, Vice President of Hanyang University, also used the image to celebrate the contributions of engineers and scientists: "Twenty years ago, we had a 'hungry spirit.' When we set a goal, we charged for it. ..."[4]

The hunger was to compete, to win. Referring to recent decades, Vice President Kwon asserted, "One of the prominent strengths of Koreans is never to lose" (cf. Chapter 2). He related a story about how Korean companies began manufacturing semiconductor memory chips. "We were bothered [by competition from] a lot by Japanese counterparts," he said. "While we were developing a product [that used their semiconductor chips], they supplied a chip at $1." "Once we successfully developed the product," however, "they raised the chip price to $5. ... [T]hey could make more money selling the finished product rather than just the chip." The result was "a strong desire to win," a hungry spirit to compete, in this case with Japan, and a thriving memory chip industry.

But things had changed. The clear message from the elite Academy was that the next generation of engineers had emerged with reduced professional capabilities, loss of the commitment to serve Korea as a whole, and, hence, a dangerous decline in individual will power. "Engineers in their thirties and forties have lost the spirit of challenge or hunger," complained Lee Hyun-soon, Vice President of Doosan Infracore, equipment maker for heavy industry.[5] "Their will power has weakened. They always try to do things in an easy way." As a result, he claimed, "[T]he competence of [our newly hired staff] in mathematics, physics, algebra is not like ours. They can't solve problems

---

[2] Cho, "Gi-Eobinui Hungry Jeongshin (Employer's Hungry Spirit)," 2007, 55. The English word "hungry" often appeared in newspapers during the middle 1960s to mean the "reality of hunger" or "reality of poverty." The term "hungry *Jeong-shin*," meaning "aggressive attitude and will to overcome poverty and difficult situations," began to appear in the 1980s. The longstanding and visible presence of the U.S. military contributed to the acceptance of English words in Korean expressions.

[3] National Academy of Engineering of Korea (NAEK), "Interviews with Engineering Leaders," 2012, 87.

[4] National Academy of Engineering of Korea (NAEK), "Interviews with Engineering Leaders," 2012, 14. Hanyang University was founded in 1939 during the Japanese Occupation, as the Dong A. College of Engineering. It was the first private engineering college in Korea. Its founders were seeking to build domestic engineering capacity to support an independent Korea.

[5] In 2012, Doosan Infracore was a subsidiary of Doosan Group, a South Korean conglomerate. See Chapter 3 for more on Korean conglomerates, the *chaebols*.

as well as we did." Added Kwon Oh-kyung, "When we [leaders] say 'Let's go!,' the youth reply 'Yes.' But they then immediately turn around and say 'Do it yourself.' It's like that now."[6]

Younger engineers and scientists tended to disagree, calling attention to relatively low pay even for those involved in industrial R&D. Consider, for example, a 2004 post to the website for the Biological Research Information Center, supported by Pohang University of Science and Technology. It lamented the life of the laboratory researcher:

> Some people regard spending sleepless nights doing research as the greatest demonstration of the merit of a [scientist-engineer]. I cannot agree with that. There are few people who are willing to do their best and invest great passion in their work without adequate compensation or regard. It is like they are squeezing every last penny out of us, even though we have already worked for years at the salary level of day laborers.[7]

If the country no longer values the work of engineers and scientists, then where are they supposed to find passion? Consider the 2011 complaint of one engineer posted to the popular scieng.net forum:

> While talking about competition in the global world, instead of nurturing us for real competitiveness they drown us with a sense of defeat. I believe that those who are obsessed with economic competitiveness should control this obsession. I know they were born in the 1940s and 1950s, when Korea was literally famished, and they achieved success, ironically, with their unwavering hungry spirit. I would like to ask: If we are no longer famished, how do you expect us to have a hungry spirit? Just because your way led you to succeed, it does not mean that is the only way to do it. Why not think of alternatives that would fit our current society better?[8]

The purpose of this book is to introduce you to what it has meant to be an engineer across the Republic of Korea, or South Korea. It does so by pursuing five basic questions about the identities, expertise, and commitments of engineers across the country: (1) How and why did Korea create new categories of technical workers, including engineers? (2) How did engineering education and engineering practices emerge? (3) Who has gained the opportunity to claim the identity of engineer and who has not? (4) How have engineers been educated and trained and where have they

---

[6] National Academy of Engineering of Korea (NAEK), "Interviews with Engineering Leaders," 2012, 126.

[7] Huks, *Hungry Jeongshin-Eun Jochi Anta!* (*Hungry Spirit Is Not Good!*), 2013. The Pohang University of Science and Technology was established in 1986 as the Pohang Institute of Science and Technology. Founded by the then-CEO of the Pohang Iron and Steel Company, POSTECH was modeled on Caltech, the California Institute of Science and Technology.

[8] LSCO, *Yo-Jeum Jeol-Meuni-Deul Mu-Eosi Munje Inga?* (*The Young People of Today, What's Wrong with Them?*), 2013. Initiating activities in 2002, *Scieng* formed as a voluntary organization of scientists and engineers dedicated to protecting the rights of scientists and engineers, participating in science and engineering policy, and promoting scientists and engineers. In 2013, it had 35,000 members.

tended to work? (5) What emerged as key issues for engineers and engineering across South Korea during the early 21st century?

In its most important sense, this book is not only *for* you but also *about* you. By helping you investigate and understand the emergence of engineers across Korea, we seek to enable you to ask more informed and sophisticated questions about yourself—including your knowledge and expertise, your commitments, and your identity.

Perhaps you consider yourself a Korean engineer. We want to help you reflect and critically analyze the specific challenges you have experienced in seeking to become an engineer and working as an engineer. Some types of questions you can readily answer: How would you describe the income and educational levels of your family? What sort of educational and work pathways did you follow that led toward engineering? What opportunities did your family provide, and what expectations did you feel?

Reading this book may help you better address other types of questions: What has made your experiences seem similar to or different from others in Korea who have sought education in engineering and identities as engineers, whether as practitioners, researchers, or educators? Have you encountered distinctive challenges as a man or woman in engineering? What do you seek to accomplish as an engineer, both for yourself and for others? Are you committed to some broader sense of service beyond personal gain? What counts as broader service or personal gain for you? We hope this book can help you identify and sort out some specific ways you have combined forms of expertise, practices, and commitments you had already gained as a person with new forms of expertise, practices, and commitments you gain along the way as an engineer.

Furthermore, we hope that, by better understanding your own formation as an engineer, you can ask more informed and sophisticated questions about others you encounter, including both engineers and non-engineers, Koreans and non-Koreans. How do they locate themselves? What do they know and how did they come to know it? What do they seek or want, and how did they acquire or develop the commitment to want it?

Perhaps you do not identify as a Korean engineer but work with engineers raised and educated in Korea. Or maybe you expect to travel to Korea and work with engineers. Once again, we want to help you better reflect on and critically analyze the specific challenges you have experienced in seeking to become and work as an engineer or non-engineer. Our main pedagogical, or teaching, strategy in this book focuses on describing the emergence of Korean engineers. We have designed this book to help you ask more sophisticated questions about engineers raised and educated in Korea by improving your ability to discern differences and commonalities among their trajectories and desires, as well as between those trajectories and desires and others originating elsewhere, such as your own. We seek to help you learn to resist the urge to homogenize, or view as one, people who were raised and educated in the Republic of Korea and call themselves engineers.

A good way to see and articulate the trajectories and desires of Korean engineers is to be able to recognize and articulate dominance among them. By dominance we mean widespread acceptance of images that distinguish, label, and characterize them. A dominant image is an image whose acceptance has scaled up sufficiently across a specific population to become given, or true, for that population.[9] Gaining a better understanding of individual engineers from Korea depends on being able to inquire intelligently into their lives and career trajectories. You accomplish this by assessing seemingly contingent and unique features of those lives and trajectories in relation to dominant forms of engineering expertise, dominant practices of engineering education and work, and dominant commitments of engineers and other people across the Republic of Korea.

We must pay attention to the fact that images engineers have of themselves may or may not overlap with images that other people have of engineers, including professionals from other fields or members of general publics. Indeed, as our brief introductory story about disagreements between elder and younger engineers and scientists suggests, engineers differ in the images they have of themselves. Images of engineers may also be in flux, gaining acceptance and scaling up, or losing it and scaling down. When we offer descriptions or make observations about Korean engineers, we try to be careful to identify whose perspectives we are describing and to call attention to changes or tensions as we see them.

By addressing the five questions above, this introductory volume maps the emergence across Korea of people who could claim the label "engineer" or to be doing "engineering" (not always the same). We highlight in particular key moments for engineers and engineering across the Korean peninsula. We inquire into the extent to which key changes in the development of engineers, the knowledge and expertise of engineers, and the commitments that engineers have to their work and their country may be linked in some way to more broadly accepted changes in the meaning of the country itself. And to the extent becoming an engineer came to be an opportunity, or privilege, we ask who has had access to this privilege and who has not.

Our hidden agenda for your learning through this book is thus not so hidden. No matter what identity you bring to reading it, we want it to help you learn to work more effectively with people who define problems differently than you do. Hopefully you would agree this is an important objective for anyone. We claim it is an especially important objective for engineering students and working engineers, both inside and outside of Korea, whose education likely places or placed highest value on mathematical problem solving based on the engineering sciences.

The first step in learning to work effectively with people who *define* problems differently is to ask informed, sophisticated questions about how they are located, what they know, and what they want.[10] This book prepares you to take that first step with people raised and educated across the territory of Korea who have taken on the identity of engineer. We revisit this issue in Chapter 7.

---

[9] Downey, *The Machine in Me*, 1998, 18–31; Downey, "What Is Engineering Studies For?," 2009.

[10] Downey, "The Engineering Cultures Syllabus as Formation Narrative," 2008.

We also intend this book to contribute to research in engineering studies,[11] engineering education research,[12] and other fields in which scholars seek practices of critical participation in the engineering education and training.[13] This study sits at the nexus of five extensive bodies of research that have examined connections between engineers and national identity. One offers detailed histories exploring the development of engineers and engineering within specific countries.[14] The second examines connections between technological developments and national identity.[15] The third views engineers and engineering as offering examples of relationships between science and national identity.[16] The fourth explores how and why connections between engineers and national identity have in all cases privileged male engineers.[17] The fifth makes visible the extent to which new technologies and associated actors, including engineers, are producing "post-national" forms of organization across the planet.[18]

This book is an account of the fraught attempts of engineers to embrace Korea as a whole. In particular, it explores tensions between images of engineering practice and identity advanced by emergent Korean engineers and their advocates and more broadly held images of technical work and technical workers. We follow how engineers and their advocates championed the material contributions of engineering as necessary to advance the country and enable it to compete with others.[19] Our account shows how a powerful nationalist movement sought to categorize novel practices of technical work and new technical practitioners as essential to national coherence and advancement. We thus describe the making of engineers across Korea as attempts at "techno-national

---

[11] See for example the International Network for Engineering Studies (www.inesweb.org), which sponsors the *Engineering Studies* journal (http://www.tandfonline.com/toc/test20/current).

[12] See for example the American Society for Engineering Education (www.asee.org), which sponsors the *Journal of Engineering Education* (http://www.asee.org/papers-and-publications/publications/jee); European Society for Engineering Education (http://www.sefi.be), which publishes the *European Journal of Engineering Education* (http://www.tandfonline.com/toc/ceee20/current - .U3obN4K7l8Q); and the Korean Society for Engineering Education (http://www.ksee.org/index.html), which publishes the *Engineering Education Magazine*.

[13] Downey, "What Is Engineering Studies For?," 2009.

[14] Lundgreen, "Engineering Education in Europe and the U.S.A., 1750–1930," 1990; Meiksins and Smith, *Engineering Labour: Technical Workers in Comparative Perspective*, 1996; Chatzis, "Introduction," 2007.

[15] Hecht, *The Radiance of France*, 1998; Hood, *Shinkansen*, 2006; Mrázek, *Engineers of Happy Land : Technology and Nationalism in a Colony*, 2002.

[16] Harrison, Carol E., and Ann Johnson, "Introduction: Science and National Identity," 2009.

[17] Hacker, *Pleasure, Power, and Technology*, 1989; Oldenziel, *Making Technology Masculine*, 1999; Canel et al., *Crossing Boundaries, Building Bridges*, 2000; Tonso, *On the Outskirts of Engineering: Learning Identity, Gender and Power Via Engineering Practice*, 2007.

[18] Nordmann, Alfred, "European Experiments," 2009.

[19] "[T]he nationalist's goal is to construct a nation that resembles and, therefore, can compete with other nations. Nations may have their own distinctive characteristics, but they resemble nothing so much as other nations" (Harrison and Johnson, "Introduction: Science and National Identity," 2009, 12.)

formation," using the European term "formation" because it refers to both education and training.[20] These efforts encountered significant resistance and ultimately produced considerable ambiguity in the identities of engineers and the meaning of engineering work.

Rather than an in-depth history, this study examines the past to better understand what emerged into the present. The importation of new technologies and subsequent internal technological developments prove important to this account because engineers and their advocates treated them as evidence engineering was indeed helping Korea advance in relation to other countries. Engineering and science do appear in close contact, but the usual hierarchy between them is reversed. Here engineers use the sciences as resources to mobilize in their country-building projects. Similar to many studies of gender in engineering, our account mainly follows men engineers and finds masculinity in their making. We do call attention to a brief period during which participation by women engineers increased dramatically, and question its implications for dominant images of gender. The concluding episodes of our study suggest that when Korean engineers began to embrace images of economic competitiveness and globalization they may have begun adding identities that challenged and complicated existing commitments to the country alone. Finally, our conclusion suggests that any efforts to participate critically in the making of Korean engineers must take account both of the dominant images of engineering that engineers themselves hold and of the fraught status of those commitments.

Because of our desire to help scholars participate more effectively in practices of Korean engineering education and training, including by formulating and attempting to scale up alternative images, we elected not to frame this study as a textbook. Yet as the volume hopefully makes clear, our primary goal is pedagogical.

WHAT IS A KOREAN ENGINEER?

The English words "engineer" and "engineering" have widely varying analogs, or rough equivalents, in different languages and across different countries and continents.[21] In one sense, this book provides a first round of answers to the question, "What is a Korean engineer?"

You should treat every bit of information and interpretation we offer as an entrée and invitation into further complexities, depending upon your interests. To help prepare you to dive more deeply into specific historical moments with informed questions about Korean engineers, we begin with a brief overview of some key categories of technical workers in late 20th century and early

---

[20] We are extending here to the realm of education Sheila Jasanoff's analysis of scientific research "co-producing" science and society simultaneously (Jasanoff, "Beyond Epistemology: Relativism and Engagement in the Politics of Science," 1996; Jasanoff, *States of Knowledge*, 2004). We also extend David Kaiser's insight that education "serves as a crucible for reproducing cultural, political, and moral values" (Kaiser, "Introduction," 2005) by showing how the making of Korean engineers helped scale up a new dominant of Korea as a country.

[21] These differences are, of course, in addition to the varying meanings in English-speaking settings as well.

21st-century South Korea. Stay with us, for the ride is a bit bumpy. Allowing this initial summary to wash over you should help you understand the more detailed accounts in the chapters that follow.

During the first thirty years following what is known in the U.S. as the Korean War (1950–1953), Korean categories of technical workers evolved from emphasizing skill-based practical work to privileging university training and work in industrial research and development (R&D).[22] Changes in categories of technical workers closely followed changes in the government's industrial policies.

The Korean government began formally categorizing technical workers in 1962, a decade after the war. Its powerful Economic Planning Board issued a classification system for the entire workforce, including technical workers. The country was beginning to build substantial capacities in industry. The Board wanted to match practices of technical training to practices of industrial work (see Table 1.1).

The classification included three levels of industrial workers, officially translated by the Board as "craftsman (기능공)," "technician (기술공)," and "engineer (기술자)."[23] The Chinese characters for craftsmen, *gi-neung-gong*, drew on the root *gi-neung*, meaning "skill." The character *gong* signified "craft." The actual craftsman or, more accurately, craftsperson was a semi-skilled production worker. Since women workers made significant contributions to the light industries of the 1960s, a significant proportion of workers who achieved this category were indeed women. A craftsperson may or may not have attended a new technical high school, but the key was that she or he had fewer than three years of experience.

The respective characters for technicians and engineers, *gi-sul-gong* and *gi-sul-ja*, drew on *gi-sul*, meaning "technology." It is important to keep in mind that, at the time, "technology" basically referred to industrial machinery. Technicians and engineers were the male technologists of industrial machinery. Technicians were generally responsible for operating imported industrial machinery, under the supervision of engineers. They had attended a technical high school and achieved three or more years of industrial experience.

Also important was using the character *ja* (자), signifying "human," to distinguish engineers. Engineers were more fully human, or educated and morally developed, than technicians. They had received some form of higher education in science or engineering. As fully developed humans, they could design, plan, and supervise the technicians and craftspersons without direct supervision themselves.

As we will see, during the 1960s, the government focused on developing varieties of light industry, making consumer goods to substitute for aid and imports. Attempting to attract more peasant men into industry, the government developed new practices of techno-national formation

---

[22] The war is commonly known in South Korea as "625" owing to its initiation on June 25. In North Korea, it is known as the "Fatherland Liberation War," and in China as the "War to Resist U.S. Aggression and Aid Korea."

[23] Economic Planning Board, *Gwa-Hag-Gi-Sul Baeg-Seo* (*Science and Technology White Paper*), 1962, 35.

at the middle and high school levels. In 1963, to take account of the educational diversity of new workers, it subdivided the category of craftsman into three new categories based on education and experience. These were the "learner," *gyeon-seub-gong* (견습공), "semi-skilled craftsman," *ban-sug-lyeon-gong* (반숙련공), and "skilled craftsman," *sug-lyeon-gong* (숙련공). All still included the character *gong*, meaning "craft." Those skilled craftspersons who had both significant technical education and three years of experience could become technicians of industrial machinery, *gi-sul-gong*. Women craftspersons tended to be learners working at lower wages, while the semi-skilled and skilled craftspersons tended to be men.

Table 1.1: Classifications of technical workers by the Economic Planning Board. Source: Ministry of Science and Technology, *Gwahakgisul Yeongam* (*Science and Technology Annals*)

Level	Year	
	1962	1963
Level I	*gi-sul-ja* (engineer)	*gi-sul-ja* (engineer)
Level II	*gi-sul-gong* (technician)	*gi-sul-gong* (technician)
Level III	*gi-neung-gong* (craftsman)	*sug-lyeon-gong* (skilled craftsman) *ban-sug-lyeon-gong* (semi-skilled craftsman) *gyeon-seub-gong* (learner)

The *gi-sul-ja*, or supervising human engineers, were few in number. Since higher technical education was a prerequisite, the sons of wealthy families had the best chance to become *gi-sul-ja*. Sons of poorer families who demonstrated high levels of achievement in school could use this as an opportunity to move up. Still, as we will see in Chapter 3, the demand for technical higher education paled in comparison to the humanities and social sciences in the 1960s because of the latter's long-term association with leadership.

In a sense, because actual craftspersons, technicians, and engineers all worked in the new arena of industrial production celebrated by the government, all did work that they and others considered "engineering." To do engineering was to work with industrial machinery. Remember this: engineering was work with industrial machinery. This association would continue into the 1980s and beyond.

During the 1960s, the scale of industrial development remained small, wages were low, and engineering was by no means high-status work. Wealthy families were not inclined to send their sons to become supervising engineers, let alone technicians of industrial machinery or craftsmen. We will see how the categories of technician and craftsman became both attractive and accessible to men, and sometimes women, from the lower classes.

By the late 1960s, the government had become interested in heavy and chemical industries, and it began to encourage domestic research and development. The newly established Ministry of Science and Technology established a classification system for R&D practitioners (Table 1.2). Not surprisingly, the ministry's leaders privileged a new site for techno-national formation: university education. They themselves held Ph.D.'s. In 1967, they introduced the category "research assistant," *yeongu josu* (연구조수), alongside that of the technician.

In 1969, the ministry introduced a category that would rise to dominance and remain so well into the 21st century, the "scientist-engineer," *gwa-hak-gi-sul-ja* (과학기술자).[24] Becoming a scientist-engineer required a man not only to graduate from a four-year university but also to gain substantial work experience in research and development and in technical planning and management. Because the country had so few researchers at the time, there was no point in attempting to distinguish scientists from engineers. Also male, the scientist-engineer was more than educated and fully human. He was understood to be a scholar and, hence, a professional who undertook creative activities in the workplace. He had a chance of passing a civil service examination and becoming a "scholar-official." In Chapter 2 we outline the longstanding significance of the scholar-official across the Korean peninsula.

With the government's new interest in heavy industries during the 1970s, including steel and heavy equipment, and chemical industries, engineering work in those industries gained higher status. New educational initiatives aimed to prepare technicians and engineers for the heavy and chemical industries.

In 1973, the Ministry of Science and Technology upgraded the category "craftsman" from *gi-neung-gong* to *gi-neung-ja*. Predominantly male, they became higher-wage skilled workers who had demonstrated humanness by completing vocational high school and gaining appropriate experience. Likewise, the technician, *gi-sul-gong*, became a field technician, *hyeon-jang-gi-sul-ja* (현장기술자). Male field technicians had become fully "human" by graduating from a new junior technical college and then producing and operating heavy machinery and more complex processes. The scientist-engineer, *gwa-hak-gi-sul-ja*, included those men who had completed master's and Ph.D. degrees, thus emphasizing their creative abilities and activities.[25] These efforts to upgrade engineering work in heavy and chemical industries culminated with the National Technical Qualification Act in 1973. Its main innovation was to multiply categories at the top.

[24] Han, "A Crisis of Identity," 2010.

[25] Ministry of Science and Technology, *Gwahakgisul Yeongam* (*Science and Technology Annals*), 1973, 72.

Table 1.2: Classifications of technical workers by the Ministry of Science and Technology and 1973 Qualification Act. Source: Ministry of Science and Technology, *Gwahakgisul Yeongam* (*Science and Technology Annals*)

Level	Year			
	1967	1969	1973	1973 Act
Level I	*gi-sul-ja* (engineer)	*gwa-hak-gi-sul-ja* (scientest-engineer)	*gwa-hak-gi-sul-ja* (scientest-engineer)	*gi-sul-sa* (professional engineer) *gi-neung-jang* (master craftsman) *gi-sa* (engineer)
Level II	*gi-sul-gong* and *yeongu josu* (technician and research assistant)	*gi-sul-gong* (technician)	*hyeon-jang-gi-sul-ja* (field technician)	*san-eob-gi-sa* (industrial technician)
Level III	*gi-neung-gong* (craftsman)	*gi-neung-gong* (craftsman)	*gi-neung-ja* (craftsman)	*gi-neung-sa* (craftsman)

Beginning in the 1980s, a new government developed and implemented policies to dramatically expand the domestic development of industrial equipment, from large machines to small electronics. The new policies produced a further upward shift in emphasis in technical education, toward university-based education and graduate schools. The longtime threefold categorization of craftsman, technician, and engineer collapsed into a twofold categorization of technicians and engineers.

During that decade, the number of students completing B.S., M.S., and Ph.D. degrees specifically in engineering increased dramatically. While roughly 11,000 students had completed B.S. degrees in engineering in 1980, by 1990 this number had increased to more than 30,000. The number of M.S. degrees increased from roughly 800 to nearly 4,000, and Ph.D.'s from 41 to 456. While the Ministry of Science and Technology began to distinguish the scientist, *gwa-hak-ja* (과학자), from the engineer, *gi-sul-ja* (기술자), the image of the scientist-engineer, *gwa-hak-gi-sul-ja* (과학기술자) in fact scaled up into most common usage across the country. It became the primary public label for engineers. It identified the accomplished techno-national scholar, a morally complete human who applied creative initiative in further developing Korean industry and might sometimes become a scholar-official.

Table 1.3: Classifications of research workers by the Ministry of Science and Technology. Source: Ministry of Science and Technology, *Gwahakgisul Yeongam* (*Science and Technology Annals*)

Level	Year	
	1980	1982
Level I	*gwa-hak-ja* (scientist) *gi-sul-ja* (engineer)	*gwa-hak-ja* (scientist) *gi-sul-ja* (engineer)
Level II	*gi-sul-gong* (technician)	*gi-neung-ja* (technician)
Level III	*gi-neung-gong* (craftsman)	*gi-neung-ja* (technician)

As we explain in Chapters 4 and 5, the rapidly expanding number of those who had completed university education in engineering began to organize and advocate on their own behalf. They advanced the term *gong-hag-in* (공학인) to label members of this exclusive group, whether they went off to work as research scientist-engineers or as technical managers.[26] The formulation was clever. Remember that *gong* refers to craft. Here, *hag* refers to an occupational area equipped with educational qualification, and *in* stands for human, as in a fully developed human. Thus, *gong-hag-in* emphasized that those who bore its label had completed higher education, had achieved higher-class status and a better occupation, and therefore deserved higher salaries.

Women students began to gain entry to this group during the 1980s. Only 150 women had earned B.S. degrees in 1980, with but 15 gaining the M.S. degree and zero completing the Ph.D. But in 1990, more than 2,000 completed the B.S., 150 the M.S., and 8 the Ph.D. After 1990, this shift continued, with even greater emphasis on the engineering graduate schools.[27]

Table 1.4: Gender composition of engineering graduates, 1980–2005. Source: Ministry of Science and Technology, *Gyoyuk Tonggye Yeonbo* (*Statistical Yearbook of Education*)

Year	B.S. Engineering		M.S. Engineering		Ph.D. Engineering	
	Total #	Total # Women	Total #	Total # Women	Total #	Total # Women
1980	11,311	150	834	15	41	-
1985	23,448	550	2,734	46	197	7
1990	30,514	2,049	3,872	150	456	8
1995	36,032	2,722	6,188	364	850	23
2000	51,673	9,176	12,513	1,024	1,538	68
2005	69,419	13,722	13,470	1,746	2,138	149

[26] We speculate that those Korean engineers who preferred *gong-hag-in* to *gi-sul-ja* wanted to distance themselves from the image of technology as industrial machinery, which still conveyed low status.

[27] Ministry of Science and Technology, *Gwahakgisul Yeongam* (*Science and Technology Annals*), 1990, 212–220. Ministry of Education, *Gyoyuk Tonggye Yeonbo* (*Statistics Yearbook of Education, Annual*).

One take-away message from all this is that, during the 30 years following 1962, successive governments shifted emphases upward in the production of engineering workers, i.e., workers in industry. The shifts paralleled changes of emphasis in industry. The government highlighted training for craftspersons from vocational middle and high schools during the light industry era of the 1960s. It turned to technicians from junior colleges during the heavy industry era of the 1970s. It privileged degreed engineers from universities and graduate schools when the government promoted more advanced electronic instruments and equipment during the 1980s. While women workers had figured prominently during the low-wage era of the light industry, they had to fight their way into higher-income categories during the later periods.

As we suggested above, during the early 2000s advocates for engineering workers argued that the status of engineering was in crisis. A government directive in 2004 sought to support engineering workers by expanding the engineering workforce beyond the popularly recognized scientist-engineers. It included all those who (1) held degrees in science or engineering, including the *gong-hag-in*; (2) passed an exam on the job developed under a new National Technical Qualification Regulation; or (3) successfully demonstrated equivalent education or training, e.g., outside the country. Understanding this concern will require understanding the continued dominance of education in the humanities and social sciences, especially law.

Also, after 2000, note that the English word "engineer" actually began to appear more frequently in popular contexts. Those who appealed to it were labeling themselves or others as ready to compete in international contexts in which the English word dominated. We address this phenomenon in Chapter 6.

In sum, by the early 21st century, the categories scientist-engineer, *gwa-hak-gi-sul-ja*; engineering graduate, *gong-hag-in*; and engineer dominated popular discourse across South Korea. Helping you better understand the trajectories of these categories is a core objective in Chapters 2–6. We show how key moments in their emergence were linked to government struggles to define and maintain South Korea as a sovereign country. Note in particular the seeming triumphs of engineers during the 1970s.

By 2010, the image of the scientist-engineer, i.e., one who does industrial research and development, had remained dominant in official circles. The government relied on it to demonstrate the country's high level of per-person commitment to industrial research. That year, the government proudly reported that 11.1 of every 1,000 workers were scientist-engineers, higher than major competitors Japan, U.S., Germany, and China.

Table 1.5: Numbers of researchers in five countries[28]				
	Korea 2010	**Japan 2010**	**U.S. 2006**	**Germany 2010**	**China 2010**
# of researchers per 1,000 employees	11.1	10.4	9.5	8.1	1.6
Total # of researchers	264,118	656,032	1,414,341	327,198	1,210,841

Numerical leadership among industrialized countries did not, however, translate into political leadership at home.

GOVERNMENT STRUGGLES TO ESTABLISH SOVEREIGNTY

We turn now to briefly map earlier governmental struggles to establish sovereignty across the Korean peninsula. Changes in leadership and governance always produce winners and losers. Some gain power; others lose it. Some become more visible, others less so. In this book we are interested in how different images of governance across the Korean peninsula scaled up and down over time, gaining and losing acceptance. We find that longstanding difficulties in establishing and maintaining sovereignty help us understand the passions in more recent governmental initiatives to formulate and attempt to scale up new images of Korean engineers.

We speak of the "Korean peninsula." Always keeping in mind the location of this territory in East Asia is essential to understanding the tensions, conflicts, and power relations it has grounded and their significance for the eventual emergence of engineers. The Korean peninsula has been contested terrain for centuries. It has been part of or occupied by powerful states, lived alongside or between powerful states, and divided between competing states.

The peninsula currently hosts two sovereign countries, the Republic of Korea, or South Korea, and the Democratic People's Republic of Korea, or North Korea (Figure 1.1). North Korea borders China to the northwest and Russia to the northeast. South Korea is separated from the southwestern islands of Japan by the Korea Strait, roughly 200 km wide at its narrowest point. Taiwan is less than 1,400 km to the south. The U.S. is roughly 9,000 km away, but it long maintained a strong military force in the region. In the early 2000s, more than 28,000 U.S. troops were in South Korea and more than 35,000 on Okinawa, Japan.

---

[28] Organization for Economic Cooperation and Development (OECD), "Main Science and Technology Indicators," 2011, 26–27.

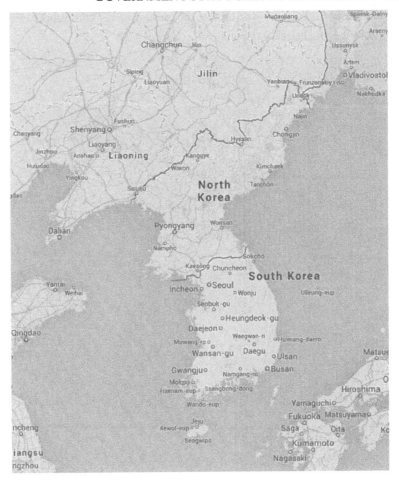

Figure 1.1: Korean peninsula, 2013. Source: Map data ©2013 AutoNavi, Google, SK planet, ZENRIN.

Scanning across more than two thousand years of Korean history, we do not begin zooming in on the emergence of engineers specifically until the late 19th century. In particular, Japan initiated a series of actions to gain control over the territory in 1876, culminating in formal colonization in 1910. The clear, evolving Japanese desire to make the territory a subordinate part of Japan plays an important part of our analysis below. Engineers began to populate the Korean peninsula after 1910, but these were not Korean engineers.

Earlier events and identities matter greatly to this account. In the first place, both the current separation between sovereign states and the earlier experience of subordination have precedence in the past. The dominant account of the peninsula's history begins with the separation of three kingdoms—Silla, Baekje, and Goguryeo—during the 1st century C.E. (Common Era in the Western calendar), along with the smaller Gaya confederacy (Figure 1.2).

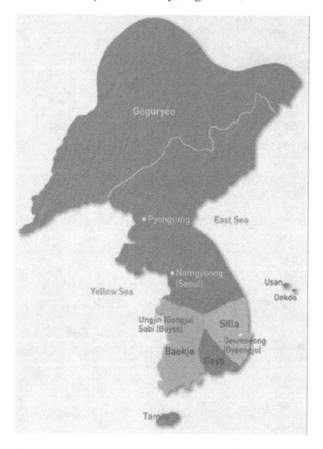

Figure 1.2: The Three Kingdoms and Gaya (5th Century). Source: www.korea.net.

The northern kingdom of Goguryeo extended beyond the peninsula into Manchuria, what is now northeastern China. By the 5th century, its capital was in Pyongyang, now the capital of North Korea. Baekje was founded near present-day Seoul, the capital of South Korea.

Silla annexed Gaya during the 5th century and united the three kingdoms under its rule in the 7th century. Established in the 10th century, Goryeo toppled Silla and ruled until Ghengis Khan conquered the peninsula in the 13th century, making Goryeo a tributary state of the Mongolian empire. After the Mongolian empire fell in the late 14th century, a Goryeo general successfully gained control of the peninsula through a coup, establishing the Joseon dynasty.

Joseon would rule the peninsula for 500 years. Although technically independent of the Chinese empire, Joseon also accepted subordination as a tributary state—at least officially, for there was much resistance. Cultural and political subordination to China brought with it a centuries-long flow of cultural influence in one direction—from China through Joseon to the Japanese archipelago. Most prominent for later Joseon identities had been the Chinese writing system in the 2nd century and Buddhism in the 4th century, followed by neo-Confucianism in the 14th century, after the Joseon dynasty had gained power.[29] With neo-Confucianism came a hierarchy of social classes whose stability and affirmation of authority via scholarship had great significance for popular assessments of engineering workers well into the 21st century.

At the top of the hierarchy were the scholar-officials, *sa* (사), mostly landed gentry, or *yangban* (양반). They demonstrated their superior virtue and qualification for elite status in ruling bureaucracies by learning to read aloud and interpret the Four Books and Five Classics of Neo-Confucian thought and practice.[30] The remaining three social classes were peasants, *nong* (농), artisans, *gong* (공), and merchants, *sang* (상). Dominant throughout the Joseon dynasty, the *Sa-nong-gong-sang* (사농공상) categories ensured that anyone involved in craftwork held a low social status.

An instructive example of Joseon desires to minimize subordination to the Chinese was creation of its own alphabet in the 15th century. Unlike the Chinese alphabet, whose characters tend to represent physical objects or abstract concepts and which require much study to master, the *Hangeul* (한글) alphabet was much simpler. It was organized phonetically, representing the sounds of the spoken language. Codifying it was an attempt to preserve it.

In 1876, the Japanese forced Joseon to accept an unequal treaty that officially ended Joseon's status as a tributary state of the *Qing* (ch'ing) dynasty. It gave extraterritorial rights to Japanese citizens and traders access to the ports of Busan, Incheon, and Wonsan. Further incursions and increasing threats from the Japanese, as well as fears of Europeans and Americans, led the Joseon king to formally declare independence from *Qing* China and establish the Great Korean Empire, *Dae-han-je-guk* (대한제국), in 1897. The Korean empire would last only 13 years.

After dramatically winning a war with Russia in 1905, Japan forced a treaty onto the Koreans that placed all trade through Korean ports under Japanese supervision and gave Japan full authority over Korea's foreign affairs. Formal annexation in 1910 initiated a 36-year attempt to make Korea wholly a subordinate province of Japan, akin to Okinawa and Hokkaido. Indeed, the Japanese government sought to eliminate all vestiges of Korean autonomy, difference, and, arguably, identity.

The end of Japanese colonization did not mark the beginning of Korean sovereignty. Partition by U.S. and Soviet military forces, occupation of the Republic of Korea by the U.S. military, temporary self-rule, a civil war, and years of dependence upon external relief kept governance ambiguous across the Korean peninsula. The 1961 rise of General and then President Park Chung-hee

---

[29] Confucius lived five hundred years before the Common Era. Developed by scholars during the 11th century, Neo-Confucianism sought to rationalize its ethical philosophy to better guide everyday life.

[30] The Four Books: *The Great Learning*; *The Doctrine of the Mean*; *The Analects*; *The Mencius*. The Five Classics: *Classic of Changes*; *Classic of Poetry*; *Classic of Rites*; *Classic of History*; *Spring and Autumn Annuals*.

initiated an 18-year effort by the South Korean government to achieve sovereignty through industrial development. It is no coincidence that the figure of the engineer emerged as an integral part.

KOREAS AND ENGINEERS

The chapters that follow trace linkages between changes in governments across the Korean peninsula and prospects for the development of engineering and engineers. Our central thesis is that what came to count as engineers across the Republic of Korea in the south was the product of explicit efforts by Park Chung-hee's government to scale up images of national progress through industrial development.[31] The Park era was a story of state-led initiatives in techno-national formation, with emergent engineers as followers taking advantage of new-found opportunities for visibility and status. In the post-Park episodes, we find both agents of the state and engineers, operating both individually and collectively, seeking ways to imagine and position engineering so its practitioners could actively embrace Korea as a whole and feel embraced by it. As the subsequent episodes illustrate, every initiative also drew inspiration, images, and resources from territories elsewhere. Each effort of a Korean government to mobilize support for engineers was an attempt to fit Korea to a contemporary world defined by some image that had scaled up to dominance. Each effort by engineers and engineering organizations to privilege their expertise and identities was an attempt not only to connect engineering to Korea but also to position Korean engineers alongside colleagues around the world.

Chapter 2, "Five Koreas without Korean Engineers: 1876–1960," examines the making of technical practitioners across the Korean peninsula beginning with the late Joseon dynasty through the years following the so-called Korean War. We identify five separate periods, listed as such, which we call the five Koreas. What ties these highly diverse periods together, for our purposes, is that at no time during any of them did a ruling government focus on producing engineers born on the Korean peninsula.

Chapter 3, "Technical Workers for Light Industry: 1961–1970," dives into the first decade of Park Chung-hee's program for creating a new Korea, focusing especially on initiatives in secondary and higher technical education. We describe the ambitious initiatives of his government during its first decade and call attention to the considerable resistance Park faced. We see that Park would succeed in producing what he called a "first economy" of light industries, although its technical workers could generally not be promoted as comparable to Western engineers. Most importantly, Park's government would not successfully scale up broad popular support, let alone acceptance, for an image of technical labor as an icon of national unity and advancement. President Park encountered huge obstacles in what he termed the "second economy."

---

[31] Downey and Lucena, "Knowledge and Professional Identity in Engineering," 2004.

Chapter 4, "Engineers for Heavy and Chemical Industries: 1970–1979," is perhaps the key chapter in the book. It documents the Park government's new emphasis on developing heavy industries, especially those producing steel and heavy equipment, as well as large-scale chemical plants. Explicit government initiatives, often through government ownership, shifted the drive for exports away from light industries and into industries that required larger-scale equipment, machinery, and manufacturing processes and, hence, higher levels of technical expertise. New engineering schools supported this effort, symbolized by banners marking their students as "Bearers of the Nation's Industrialization." Many scientist-engineers gained the opportunity to serve as scholar-officials in the Park government. Growth in the heavy and chemical industries elevated the country to the status of an international economic power.

Yet government-led economic development depended on increasing inequality and overt repression, justified by claims of national emergency. Systematically repressing dissent in the name of prosperity would not substantially upgrade the status of vocational secondary education, technical higher education, and technical practitioners at all levels, including scientist-engineers. In fact, overt repression became one indicator that the governmental vision of techno-national formation and advancement was not achieving universal support.

Chapter 5, "Loss of Privilege and Visibility: 1980–1998," follows the decline of Park's program to elevate technical workers, especially engineers, to the status of scholar-officials. New images of Korea and Korean advancement relative to other countries scaled up to dominance. During the 1980s, demands for democratization contributed to shifting leadership in industrial development away from government and toward the private conglomerates, the *chaebols*. By the early 1990s, further democratization, understood through the new lens of economic competitiveness, came to mean highlighting individual opportunity over collective responsibility. The government's longtime focus on heavy and chemical industries gave way to more diverse private initiatives in electronic, information, and communication technologies. With a loss in official advocacy, the practitioners of engineering, from scientist-engineers to technicians, lost coherence and visibility as icons of the country's advancement.

Chapter 6, "Engineers for a Post-Catch-Up Korea," examines the emergence of and responses to what engineers and their advocates came to call the "science and engineering crisis." It was a crisis of identity. Late-career scientist-engineers and their advocates continued to campaign for the government to restore their elite status as technical scholar-officials. Although engineering organizations received some support through legislation and governmental policies, the support was uneven and inconsistent. We then turn to explore the continuing struggles of women engineers, continuing association of engineering with military hierarchy and discipline, and new images scaling up to challenge early-career men and women engineers. Once again we find that engineers, scientist-engineers, and other technical workers appeared to have few resources at their disposal to restore their status to the level of scholar-officials. This did not, however, stop them from trying.

Chapter 7 reviews our answers to the five questions that framed our analysis and defined its learning objectives for different readers. It provides an overview of the struggles of the highest-ranked Korean engineers to claim the identity of scholar-official and thereby embrace Korea as a whole. We then turn to question how understanding attempts to produce techno-national formation for Korean engineers might help engineering students and working engineers both inside and outside of Korea better reflect critically on their identities, expertise, and commitments. Finally, we briefly examine some of the challenges this study identifies to anyone, including scholars, who might seek to scale up alternatives to dominant images of engineers and engineering across Korea.

REFERENCES

Canel, Annie, Ruth Oldenziel and Karin Zachmann. *Crossing Boundaries, Building Bridges: Comparing the History of Women Engineers*, 1870s-1990s. Amsterdam, Abingdon: Harwood Academic; Marston, 2000. 6

Chatzis, Konstantinos. "Introduction: National Identities of Engineers." *History and Technology* 23, no. 3 (2007): 193–196. DOI: 10.1080/07341510701300239. 6

Cho, Jung-shik. "Gi-Eobinui Hungry Jeongshin (Employer's Hungry Spirit)." <manuscript>, 2007. 2

Downey, Gary Lee. *The Machine in Me: An Anthropologist Sits among Computer Engineers*. New York: Routledge, 1998. DOI: 10.1525/ae.2000.27.1.182. 5

Downey, Gary Lee. "The Engineering Cultures Syllabus as Formation Narrative: Critical Participation in Engineering Education through Problem Definition." *St. Thomas Law Journal* (special symposium issue on professional identity in law, medicine, and engineering) 5, no. 2 (2008): 101-130. 5

Downey, Gary Lee. "What Is Engineering Studies For?: Dominant Practices and Scalable Scholarship." *Engineering Studies: Journal of the International Network for Engineering Studies* 1, no. 1 (2009): 55-76. DOI: 10.1080/19378620902786499. 6

Downey, Gary Lee and Juan Lucena. "Knowledge and Professional Identity in Engineering: Code-Switching and the Metrics of Progress." *History and Technology* 20, no. 4 (2004): 393-420. DOI: 10.1080/0734151042000304358. 18

Economic Planning Board. *Gwa-Hag-Gi-Sul Baeg-Seo (Science and Technology White Paper)*. Seoul: Economic Planning Board, 1962. 8

Hacker, Sally. *Pleasure, Power, and Technology: Some Tales of Gender, Engineering, and the Cooperative Workplace*. Boston: Unwin Hyman, 1989. 6

Han, Kyonghee. "A Crisis of Identity: The Kwa-Hak-Ki-Sul-Ja (Scientist-Engineer) in Contemporary Korea." *Engineering Studies* 2, no. 2 (2010): 125-147. DOI: 10.1080/19378629.2010.490557. 10

Harrison, Carol E. and Ann Johnson. "Introduction: Science and National Identity." *Osiris* 24, no. 1 (2009): 1-14. DOI: 10.1086/605966. 6

Hecht, Gabrielle. *The Radiance of France: Nuclear Power and National Identity after World War II.* Cambridge: The MIT Press, 1998. 6

Hood, Christopher P. *Shinkansen: From Bullet Train to Symbol of Modern Japan.* London: Routledge, 2006. DOI: 10.1215/s12280-008-9034-9. 6

Huks. *Hungry Jeongshin-Eun Jochi Anta! (Hungry Spirit Is Not Good!).* (http://bric.postech.ac.kr) Pohang University of Science and Technology, Accessed February 15, 2013. 3

Jasanoff, Sheila. "Beyond Epistemology: Relativism and Engagement in the Politics of Science." *Social Studies of Science* 26, no. 2, Special Issue on 'The Politics of SSK: Neutrality, Commitment and beyond' (1996): 393-418. DOI: 10.1177/030631296026002008. 7

Jasanoff, Sheila, ed. *States of Knowledge: The Co-Production of Science and Social Order.* London: Routledge, 2004. DOI: 10.4324/9780203413845. 7

Kaiser, David. "Introduction: Moving Pedagogy from the Periphery to the Center." In *Pedagogy and the Practice of Science: Historical and Contemporary Perspectives*, edited by Kaiser, David, Cambridge, Massachusetts: MIT Press, 2005. DOI:10.1017/S0007087406419373. 7

Korea Federation of Science and Technology Societies (KOFST). "Declaration of Crisis in Science and Technology." <manuscript>, 2002. 1

LSCO. *Yo-Jeum Jeol-Meuni-Deul Mu-Eosi Munje Inga? (The Young People of Today, What's Wrong with Them?).* [LSCO 2011.5.6 http://www.scieng.net] Scieng, Accessed February 25, 2013. 3

Lundgreen, Peter. "Engineering Education in Europe and the U.S.A., 1750-1930: The Rise to Dominance of School Culture and the Engineering Profession." *Annals of Science* 47, no. (1990): 33-75. 6

Meiksins, Peter and Chris Smith. *Engineering Labour: Technical Workers in Comparative Perspective.* London and New York: Verso, 1996. 6

Ministry of Education, *Gyoyuk Tonggye Yeonbo (Statistics Yearbook of Education, Annual)*, Seoul: 12

Ministry of Science and Technology. *Gwahakgisul Yeongam (Science and Technology Annals).* Seoul: MOST, 1973. 10

Ministry of Science and Technology, "*Gwahakgisul Yeongam (Science and Technology Annals)*," Seoul: 1990. 12

Mrázek, Rudolf. *Engineers of Happy Land : Technology and Nationalism in a Colony*. Princeton, N.J.: Princeton University Press, 2002. 6

National Academy of Engineering of Korea (NAEK), "Interviews with Engineering Leaders " Seoul: NAEK, 2012. 2, 3

Nordmann, Alfred. "European Experiments." *Osiris* 24, no. 1 (2009): 278-302. DOI: 10.1086/605985. 6

Oldenziel, Ruth. *Making Technology Masculine: Men, Women and Moderen Machines in America* 1870-1940. Amsterdam: Amsterdam University Press, 1999. DOI: 10.5117/9789053563816. 6

Organisation for Economic Cooperation and Development (OECD), "Main Science and Technology Indicators, 2010," Paris: Organisation for Economic Cooperation and Development (OECD), 2011. 14

Tonso, Karen. *On the Outskirts of Engineering: Learning Identity, Gender and Power Via Engineering Practice*. Rotterdam: Sense Publishers, 2007. DOI: 10.1177/ 0162243908321281. 6

CHAPTER 2

Five Koreas Without Korean Engineers: 1876–1960

"How lazy we are," wrote the Joseon reformist Jeong Yak-yong (1762–1836) in 1802. Later known by the penname Dasan, meaning "mountain of tea," he was criticizing disinterest in machines and manufacturing across Joseon. "The hundreds of techniques and artisan skills in our country were learned from China in the past," he asserted, "but for hundreds of years we have not developed a plan to learn more, like doing so was censored." He pointed out that in China "new methods and elaborate manufacturing technologies to make machines are increasing daily." In contrast, "we don't discuss this and feel content with the old methods."[1]

Note when Dasan was writing. Subjects of the Joseon (*Yi* family) dynasty at that time were still looking primarily toward China for insight and direction. Traders from Portugal, the Netherlands, Great Britain, and France were in China, but they still focused on acquiring such goods as porcelain, silk, spices, and tea. Rapid growth of industry in Britain had not yet translated into the desire to ensure stable access to natural resources and markets by actively controlling East Asian territory.

Dasan had passed the literary exam that gained him admission to the national Confucian academy, and he had finished first in the higher civil service exam. This gained him a position in the Office of Royal Decrees and aristocratic status as a scholar-official. With an unusual interest in construction, he had designed a pontoon bridge and both designed and supervised the construction of walls for a fortress (Figure 2.1). Although he proposed that the Joseon government establish a governmental office to manage the introduction of technologies for industry and national defense, this did not happen.[2]

Much higher on the agendas of the elite, educated bureaucratic officials in the early 1800s than introducing industry was stopping or containing the invasion of Catholicism. This Christian religion directed faith and duty away from the king. Joseon scholar-officials and, more broadly, rural *yangban* aristocrats feared it would undermine respect for ancestors and encourage resistance among peasants to their rule.

Although Dasan was reportedly not a believer in Christianity himself, members of his family were executed for beliefs that undermined loyalty and filial piety. The dominant image of filial piety in Neo-Confucianism as it was practiced across the Korean peninsula specified a flow of respon-

---

[1] Kim, "Geundae Hangugui Gisul Gae-Nyeom (The Concept of Technology in Modern Korea)," 2012, 320.

[2] Kim, "Jeong Dasanui Gwahak-Gisul Sasang (Jeong Dasan's Thought on Science and Technology)," 1989, 288.

sibilities to oneself, one's family, one's superiors, and ultimately the king. Spending years in exile, Dasan developed and sought to scale up a new approach to Confucianism. It added a commitment to collective material progress via technical work to Confucian challenges to produce a more righteous government and more moral, egalitarian society.[3] Depending on your point of view, Dasan was either far ahead of his time or dramatically out of sync.

 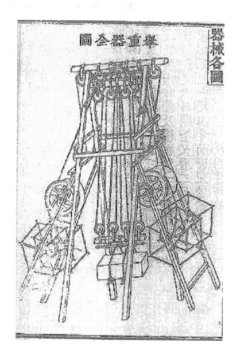

Figure 2.1: Dasan Jeong Yak-yong and the pulley (*Geojunggi*) he designed to construct Hwaseong Fortress. Source: Jeollanam-do Gangjin-gun.

In this chapter, we examine the making of technical practitioners across the Korean peninsula from the late Joseon dynasty to the years following the so-called Korean War. We identify five separate periods, which we call the five Koreas. What ties these highly diverse periods together is that at no time during any of them did a ruling government focus on producing engineers born on the Korean peninsula.

The chapter begins by describing late Joseon efforts to develop capacity in industry under pressure from the Japanese. It explains how Japanese colonialists limited opportunities for technical education to preparation as low-status technicians. They were certain that Koreans were racially inferior beings.

---

[3] Song, "Jeong Yak-Yong's Thoughts on Technology," 1994, 275–276.

We then examine the limited interest in engineering education and engineers after the Pacific War. We look first at the three-year military occupation by the U.S. and then follow the periods both before and after the war between the People's Republic of Korea to the north and the Republic of Korea to the south.

LATE JOSEON DISINTEREST IN PHYSICAL LABOR: 1876–1897

Until the Japanese empire forced the Joseon government to accept an unequal treaty in 1876, only small numbers of literati, members of the educated class, had produced books about European learning or even sought access to it. The geographical pathway for this flow was through China, to which the Joseon government dispatched scholarly emissaries on an annual basis. In most cases, the curiosity was scholarly in content and the reaction was surprise.

The elite male emissaries produced reports (*Yeon-Haeng-Rok* Selection) describing and assessing what they saw. Some also produced personal diaries. The diary of Kim Chang-eop (1658–1721), for example, describes his visit to a Jesuit cathedral (*Cheon-ju-dang*) and observatory (*Chun-mun-dae*) in the Ch'ing capital (Beijing) in 1713. He expressed fascination at the celestial globe and alarm clock at the observatory but could not explain what they were.[4]

Joseon scholar-officials from the *yangban* (양반) aristocracy had adequate intellectual resources at their disposal for explaining the active responsibilities of humankind. These included building harmony between the individual person and a universe that was not fundamentally separate or distinct. Images and practices of engaging this universe were plentiful. The childhood and adolescent work of memorizing and appealing to Confucian texts charged each person to purify one's "way (도)." Because one's way as a moral journey was necessarily encased in the material body, one had to engage matter to purify oneself. And engaging matter included, for example, "*li-yong* (이용)," meaning "facilitating the world" or the act of "plotting life's proceeds."[5]

Joseon scholar-officials also had available to them other concepts even closer to our interest in the emergence of engineers and engineering. For example, "*gae-mul* (개물)" named the active "development of nature" or the "synthetic manufacturing of nature."[6] In contrast with facilitating the world or plotting life's proceeds, actively developing or synthetically manufacturing nature stood far from the pursuit of moral value that defined humanness. Practices of *gae-mul* carried low status. Understanding this helps explain why men involved in physical labor did not carry the

---

[4] Jang, "Joseonhugi Yeonhaeng-Rokui Cheonjudang Gyeonmunroggwa Seohak Insik (An Envoy's Observational Record of the Catholic Church in Late Joseon and a Discussion of Western Learning)," 2009, 82. Since the 1580s, Jesuits in China "had accommodated differences between European and Chinese learning by focusing on mathematics and astronomy, . . . precisely because they recognized that literati and emperors were interested in such fields" (Elman, *A Cultural History of Modern Science in China*, 2006, 1–2.).

[5] Kim, "Geundae Hangugui Gisul Gae-Nyeom (The Concept of Technology in Modern Korea)," 2012, 312.

[6] Song, "Cheon-Gong-Gae-Mul (Heaven-Artisan-Open-Matter)," 2009, 430–431; Seong, *Gisul-Ui Yihae Geurigo Hangug-Ui Gisul (The Understanding of Technology, and Korean Technology)*, 1986, 66.

elevated status held by those men of comparable leisure who memorized and recited classic texts. Physical labor itself did not constitute the pursuit of moral value, and hence was secondary and inferior to mental labor. Accordingly, artisans (*gong*) who produced useful, everyday objects fell into a lower class of humans. They were higher than crass merchants but relatively removed from activities that purified one's way.

While members of the lower classes did not have access to the Chinese language and literature, literacy in Korean was remarkably high. One foreign observer, a French naval officer, observed in the 1870s:

> It is very hard to see those who are illiterate in Joseon; they are easily disdained by others for being illiterate. It seems that at least a fundamental level of education is widely distributed. Even a poor family has papers and books at home.[7]

Like their Chinese counterparts, Joseon scholar-officials had little use for the astronomy of Copernicus (1473–1543). To them, the shape of the earth and its movements were largely indifferent to principles challenging human society. In the 1700s, Hong Dae-yong (1731–1783) elaborated a theory of day and night caused by the earth's rotation (*Ji-jeon-seol*). Although well-known across both Ch'ing China to the west and the Tokugawa shogunate (Japan) to the east, the theory produced little stir beyond intellectual ruminations.[8] By the mid-1700s, populations across Europe increasingly took it as given that a God-created humanity was both separated from a mechanical nature and empowered to seek and exert mastery over it. Such was not the case across the agrarian Korean peninsula and its neighboring territories.

Writing in the 1800s, Jeong Yak-yong, described above, stood out in critiquing Confucian images of nature and working to free a logic of nature from the theoretical reasoning and intuitive rules of humankind. However, his attempts to build a literary foundation for perceiving and engaging nature in what EuroAmericans might have called a more "objective" way would be acknowledged and celebrated only in retrospect. During his time and for many years afterward, actively developing or synthetically manufacturing nature were known and understood as subordinate, pedestrian activities in a world that rewarded the spiritual, purifying process of memorizing and attempting to follow the prescriptions of sacred texts. Neither they nor their practitioners, male or female, merited significant attention from elite, male administrative leaders.

RESPONDING TO THE THREAT FROM JAPAN: 1897–1910

As we described in Chapter 1, King Gojong (1852–1919) established the Great Korean Empire in 1897. It was part of an ill-fated effort to emulate what the Japanese had done, appropriating

---

[7] Boulesteix, *Chakan Migae-in Dongyangui Hyeonja* (*Good Savage and the Oriental Wise*), 2001, 116.

[8] Park, *Hankugsa-Edo Kwahak-Eun Inneunga?* (*Does Science Exist in Korean History?*), 2004, 167; Park, *Science and Technology in Korean History*, 2005, 151.

strength from Western countries by developing industry and expanding the army. Here we examine responses by Joseon scholar-officials to increased Japanese incursions before Japan annexed the entire peninsula in 1910. The Japanese military that threatened invasion in 1875 had arrived to assert Japanese moral and racial superiority. In planetary terms, it also represented a new image of empire.

The Japanese shogunate had been a warlord federation of roughly 260 diverse and hierarchically organized domains led by the Tokugawa household. One way it maintained control over subordinate households and their domains was by limiting contact and trade with outsiders to the port of Nagasaki. A few thousand Chinese traders worked there along with a few dozen Dutch.

The shogunate government had long considered Europeans to be barbarians—uncivilized people—evidenced especially by their obsession with material wealth. Shogunate officials engaged the world through a geographic image of virtue and civilization, affirmed by military strength. The archipelago had its own version of the neo-Confucian class distinctions. Elite status was held by a male scholar who was also a warrior—*samurai*.

Joseon did have military officers, trained in martial arts, but they were lower in status than the scholar-officials. Where the civil service exam was reserved for aristocrats educated in neo-Confucian literature, male members of the lower classes who sought upward mobility tended toward the military service exam. Martial arts were not central to preparation for civil service.[9]

In contrast with China and Joseon, moral superiority and military strength across the shogunate came hand-in-hand. In addition, moral superiority and military force depended on one's geographical location. The further away one's origin was from Edo, and more generally the archipelago, the less orderly, proper, and virtuous one was. Tokugawa officials applied this calculus to the inhabitants of the Korean peninsula.[10] And by the time the Americans arrived in 1853, they were applying it to Imperial China.

EuroAmerican countries possessed obvious military power. Interpreting them through a dominant image of strength and virtue meant that the EuroAmericans must not be barbarians, but were evidently superior in moral terms as well. The new Meiji government, established in 1868, dramatically increased the Emperor's authority and responsibilities by adding practices of governance to his existing identity as a spiritual leader. Formally under his authority, government officials went to work building heavy industry in order to overcome the empire's military and moral inferiority to EuroAmerican countries. This promised the achievement of civilization and enlightenment. Cleverly, they imported European workers from whom Japanese counterparts could quickly learn practices of elevated civilization. Then, when the imported workers were no longer needed, they were sent home.

---

[9]   Cha, "Joseon Sidae Mungwa Yeongu-Ui Donghyang-Gwa Jeonmang (Trends and Prospects About Mungwa in Joseon Period Research)," 2012, 312.

[10]   The Japanese had invaded the Korean peninsula between 1592 and 1598, prior to the rise of the Tokugawa household. The invasions were unsuccessful.

For our purposes, the initiative that stands out most prominently began in 1871. Administrative leaders decided to invest a huge proportion of available tax funds into a school that would produce an entirely new category of scholar. Its all-male graduates would become sufficiently knowledgeable about industry for Westerners to call them "engineers."[11] During the 1870s and 1880s, many former *samurai* (especially mid-level in status) sought to become engineers and contribute to achieving civilization for the new Japanese empire. They committed themselves to the government's grand goal: gaining the respect of the West.

The Japanese empire was initially small and lacking in natural resources. Military officials, with support from others both inside and outside the government, called for Japan to seek and achieve hegemony in East Asia. Japan may not (yet) be superior to Europeans and Americans, so the thinking went, but it was certainly already superior across the region beyond the archipelago. The obvious place to start exerting authority and demonstrating local superiority was the Korean peninsula.[12]

Joseon responses to Japanese pressure were active but short-lived. The Great Korean Empire of 1897 was the first official step, necessary because it officially shifted power and authority away from China to the peninsula itself. Three years earlier, the Joseon government had already constructed the peninsula's first telegraph and begun sending teachers to Japan to become familiar with the practices of industry. After the King made himself emperor, the government installed the first electric lights in his palace (*Gyeong-bok*). The first telephone followed in 1898 (Figure 2.2), tram in 1898, and railroad line (19-mile line from Seoul to Incheon) in 1899.

A reporter for *The Independent* (*Tongnip Sinmun*), a cheerleader for those initiatives, dramatized an experience that reduced a twelve-hour trek to one hour and forty minutes. "When I looked outside from the carriage," the reporter wrote, "everything around the carriage appeared to be running together, and even flying birds could not catch up with it."[13]

The new Korean empire also began promoting domestic technical education that year through industrial schools for boys who were lower-level aristocrats or even commoners. It established the National Agricultural, Commerce, and Technical School to train industry leaders, as well as numerous postal schools, medical schools, and schools for miners. In addition, many private schools appeared to provide training in tanning, weaving, railroad works, and surveying. Many of these schools were established by missionaries who seized on Koreans' fears of other groups of powerful foreigners.

---

[11] Wada, "Engineering Education and the Spirit of Samurai at the Imperial College of Engineering in Tokyo, 1871–1886," 2008.

[12] Downey and Wada, "Avoiding Inferiority," 2011,

[13] Park, *Maehogui Jilju Geudae-Ui Hoengdan* (*Railway, Fashioning Run, Crossing Modern Time*), 2003, 19.

Figure 2.2: The first telephone exchange in Korea. Source: The City History Compilation Committee of Seoul.

Sixty years earlier, French priests had been caught on the Korean peninsula and executed.[14] By the 1890s, however, the military power of EuroAmerican countries and empires and the empire of Japan had far supplanted Christianity as the principal external threat. Missionaries took particular advantage of the government's official interest in industry. They proselytized Christianity while providing education for industry. By 1905, the telegraph sector had nearly 125 technicians, all male and mostly commoners.[15] By 1908, Korea had more than 100 private schools teaching surveying alone.

One resource Korean empire officials mobilized to justify its initiatives was a neologism, a new word granting status to industrial work. As early as 1890, the word *gi-sul* (기술), had traveled from Japan and was flowing across the Korean peninsula and China. To Korean scholar-officials,

---

[14] Hyeon, "1880 Nyeondae Huban–1890 Nyeondae Jeonbangi Joseonui Peurangseu Insik (French Understanding of Joseon between the Latter 1880s and the Early 1890s)," 2012, 446.

[15] Kim, "Daehanjeguggi Saeroun Gisulgwanwon Jibdanui Hyeongseonggwa Haeche (The Establishment, Development, and Disbanding of Telegraph Technicians in Imperial Korea)," 2008, 205.

the character "*gi* (기)" meant the "skill of doing industrial work" and "*sul* (술)" referred to a "method that many people follow."[16] *Gi-sul* was, in principle, a translation of the word "technology," credited to the Japanese educator Nishe Amane (1829–1879).[17] For Empire officials, accepting and using the word became a promotion. Using it was an argument for adding industry to existing moral commitments and adding industrial work to existing personal identities.

A common concern across the territory was precisely that building industry necessarily required proximity to physical labor, not to mention active participation in it. Would male *yangban* aristocrats reject industrial work as far from the elevated literary practices that purified one's moral journey? Would it pose a risk to their status? At the same time, would male peasants reject it because it failed to offer pathways to higher status? Perhaps presenting industry as a new type of work that many people followed, the meaning of *gi-sul*, could actually encourage many men, and possibly some women, to follow it?

Although competing with Japan, Joseon/Korean scholar-officials by and large did not replicate its explicit and total embrace of EuroAmerican industrial practices. It is tempting to argue that, even if they wanted to do so, their moral and administrative strategies did not carry the military force that Meiji leaders possessed in Japan. *Yangban* authority had long relied on so-called neo-Confucian images of differential superiority through literary purification. Integrating Western technology in the midst of such images of the relationship between the individual and the universe would be no easy task. Not surprisingly, different scholar-officials nominated distinct pathways, and much disagreement and conflict resulted. An especially difficult question: how could the practices of industry be justified as in some sense advancement or civilization given the known crass behavior of EuroAmericans?

The Joseon scholar Kim Yoon-shik (1835–1922) had addressed this question, for example, in 1882. He began by granting the widely held view that Western education "is wicked and similar to erotic sounds and lavishly-decorated women." Accordingly, it "needs to be ostracized." At the same time, while "we should distance ourselves" from their educational practices, he asserted that "their technology is useful."[18]

Kim offered two reasons for accepting the new and useful technology. Dominant Confucian ideals of the *yangban* literati extrapolated the proper exercise of hierarchical authority by virtuous leaders into benefits for all humankind. While building industry and infrastructures potentially confused accepted hierarchies of virtue, these new activities did offer direct approaches to collective benefit. "If we can promote public welfare," Kim continued, "why should we avoid it when making policies for agriculture, sericulture, medicine, military, ship-building, and wagons?" Kim's second

---

[16] Seong, *Gisul-Ui Yihae Geurigo Hangug-Ui Gisul* (*The Understanding of Technology, and Korean Technology*), 1986, 64.

[17] Kim, "Geundae Hangugui Gisul Gae-Nyeom (The Concept of Technology in Modern Korea)," 2012, 313.

[18] Yi, "Kim Yun-Shikui Gaehwa-Jagang-Ron-Kwa Yeongsunsa Sahaeng (Kim Yun-Sik's Enlightenment Thought and Yeongsunsa)," 2006, 100.

reason was the obvious external military threat. Joseon clearly needed greater military strength to survive. "Especially when it is apparent who is strong and who is weak," Kim asserted, "if we do not accept their technology, how do we endure their abuse and protect ourselves from their scrutiny?"[19]

A chorus of official voices agreed. During the 1890s, scholar-officials who supported appropriating practices and strength from Euro-American countries advanced the phrase "*dong-do-seo-gi-lon* (동도서기론)" to name their commitment to the new "meticulous" work. Yes, the new technical work involved physical labor, but it contrasted with the low-status practices of artisans. Akin to the exceedingly difficult work of memorizing classic texts, it required fine attention to detail. Advocates of *gi-sul* thus sought to promote the skills involved in the new meticulous work in terms of, rather than as alternatives to, familiar or established moral commitments. Their strategy: accept the meticulous work and its products but attempt to bracket these, or keep them separate, from Western images of humanity attempting to progress by mastering nature.

Official endorsement of technical education and the first steps to implement it thus did not translate into a strong desire to follow Japan and attempt to produce engineers as technical leaders. Furthermore, government promotion did not scale up into a broad, wholehearted embrace for work with industrial machinery. The official embrace met with resistance and, sometimes, conflict.[20] In 1905, the year the Great Korean Empire became a protectorate of the Japanese empire, both the National Agricultural, Commerce, and Technical School and the National Medical School had shortfalls of applicants. Some educated scholars abroad campaigned for the importance of technical education, yet few seemed to want it for themselves.

Some members of the middle class or fallen members of the *yangban* moved to acquire the new forms of meticulous learning. Yet even when they did so, they frequently had difficulty finding employment that made use of this new knowledge, let alone grant them social recognition for the effort. Disincentives to taking on the new meticulous work remained powerful.

LOW-LEVEL TECHNICIANS FOR THE JAPANESE EMPIRE: 1910–1945

It was quickly clear after 1910 that the Japanese colonial government, headed by a Governor-General, had no intention of continuing higher education to produce leaders for industry. Its 1911 Ordinance on Joseon Education required all schools to teach in Japanese and to gain approval from the Governor-General for all regulations and curricula.[21] It is instructive that the occupying government revived the name Joseon to label the territory. The old Joseon had officially deferred

[19] Yi, "Kim Yun-Shikui Gaehwa-Jagang-Ron-Kwa Yeongsunsa Sahaeng (Kim Yun-Sik's Enlightenment Thought and Yeongsunsa)," 2006, 100.

[20] Kim, "Geundae Hangugui Gisul Gae-Nyeom (The Concept of Technology in Modern Korea)," 2012, 308.

[21] Kim, "Joseon Chongdogbu-Ui Gyoyuk Jeong Chaeggwa Gyogwaseo Balhaeng (Educational Policy and the Publication of Textbooks by the Joseon Government-General)," 2009, 306.

to Qing China but had managed to remain effectively independent. This new Joseon was clearly a subordinate territory of Imperial Japan.

As the new government populated Korean offices with Japanese officials, it began either closing down or dramatically reducing the scope of both government-run and private technical schools. For example, Japanese officials reduced the curriculum at the National Agricultural, Commerce and Technical School from four years to two and renamed it the National Technical Training Center.[22] The Governor-General's office also instituted a strict restriction on study abroad. It was making clear that, in the area of technical education on the Korean peninsula, the sole purpose of formal study was to produce technicians for low-level work.[23]

Remember the Japanese image of superiority calculated as a function of distance from the center. To the occupiers now in control, native Koreans were biologically, i.e., racially, unfit for higher technical education. Sidehara Taira (1870–1953), the Japanese official in charge of education on the peninsula, asserted that Koreans aged 23 to 33, the most vigorous age group for work in industry, did not equal second- or third-year Japanese high school students in their capacities for quantitative reasoning and calculation.[24] In this view, even elite *yangban* who had memorized sacred texts lacked the mental powers to participate in the new, elevated industrial civilization. In the dominant Japanese view, how could racially inferior people on the Korean peninsula possibly be attracted to or have any use for higher education, let alone higher technical education?

A researcher in the new educational bureaucracy, Yuge Kotaro (b. 1881), candidly shared the additional justification that everyone affected by educational policies understood. In the 1923 book *Joseon's Education*, he argued that any higher technical education for Koreans should be limited to relatively elementary courses. The rationale was explicitly political. Limiting education in the new Joseon would prevent colonists from using the technical capabilities of higher civilization to revive Korean identities and national self-awareness.[25]

Under the Japanese regime, even those men who gained low-level technical training faced uncertain and insecure prospects for employment. Consider, for example, Japanese initiatives to acquire and prepare land for colonial government buildings and projects. These required considerable labor skilled in surveying, and not enough Japanese engineers were present to do the work. Also, some Joseon colonists possessed detailed knowledge about the local geological and environmental characteristics of their regions. A prominent, and repeated, Japanese strategy was to offer six-month or one-year courses of study in surveying work through a local training center. The curriculum would prepare students for temporary work on a specific local government project. When the

[22] Lee, *Hangug Gisul Gyoyuksa* (*Korean History of Technology Education*), 1991, 225.

[23] Kim, *Hanguk Geundae Gwahak-Gisul-Illyeogui Chul-Hyeon* (*The Emergence of Korean Science-Technology Personnel*), 2005, 449.

[24] Ministry of Education, "Gwahaggisul 40 Nyeonsa (Forty-Year History of Korean Science and Technology)," 2008, 29.

[25] Park, *Hankugsa-Edo Kwahak-Eun Inneunga?* (*Does Science Exist in Korean History?*), 2004, 226.

project was complete, the colonial officials in charge would then simply, and abruptly, terminate the newly skilled technicians.[26]

During the early years of the occupation, the Governor-General's office did permit some initiatives in mid-level technical education. In 1916, for example, it established the three-year *Kyeong-seong* Higher Engineering School. The school's departments included textiles, applied chemistry, civil engineering, architectural engineering, mechanical engineering, and electrical engineering.

These department names should not deceive us, however. They were not indicators of techno-national formation. Remember from Chapter 1 that as late as the 1960s industrial workers from craftsmen on upward could be counted as doing engineering work. Similarly, what the Japanese occupiers meant by engineering for Joseon in the 1910s was not elite training in science-based calculation to steer industrial development, as was the case in Japan itself. Rather, training in a higher engineering school meant technical training just high enough in its mathematical contents to prepare students for low-level positions in local public offices and as collaborating technicians and low-level managers in factories and construction sites.[27] The small departments were geared to "train technicians and managers for Joseon's industrial development," not to teach "high-level scientific principles."[28]

By 1918, eight years into the colonial occupation, Joseon had a total of 15 commerce schools, 33 agricultural schools, and 68 common industrial schools.[29] The vast majority of these offered specialized training for low-level technical work (see Figure 2.3).

It is important to understand that many members of the old *yangban* elite actively collaborated with the Japanese. Collaboration enabled them to retain ownership of land and considerable local political power.[30] It is also important to understand that many Koreans, including elites and non-elites, resisted wholesale incorporation into Japan in a variety of ways. The scope of their resistance in fact expanded over the course of the thirty-six-year occupation.

Early resistance solidified in the so-called March 1 Movement of 1919. The Movement formally began when a collection of 33 cultural leaders organized a mass demonstration on the commemoration day for the late emperor. They read a proclamation of Korean independence. During the following year, similar demonstrations took place across the peninsula. Their scale overwhelmed unprepared local forces of Japanese police. Movement leaders had to flee to China,

---

[26] Kim, *Hanguk Geundae Gwahak-Gisul-Illyeogui Chul-Hyeon* (*The Emergence of Korean Science-Technology Personnel*), 2005, 122.

[27] Of the 58 Korean graduates from the *Kyeong-seoung* Higher Engineering School's textile department between 1933 and 1939, 27 went to work in industry and 20 were hired by local public offices (Seo, "Geundaejeok Myeonbangjik Gongjang-Ui Deungjang-Gwa Gisul-Illyeok Yangseongjedo-Ui Hyeongsung (Emergence of the Large-Scale Cotton Textile Factory and Formation of a Training System for Engineers)," 2011, 116.

[28] Lee, *Hangug Gisul Gyoyuksa* (*Korean History of Technology Education*), 1991, 254.

[29] Lee, *Hangug Gisul Gyoyuksa* (*Korean History of Technology Education*), 1991, 246–247.

[30] Lie, *Han Unbound*, 1998, 181.

however, when the Governor-General called in support from the army and navy. The combined forces of police and military brutally suppressed the resistance, killing more than 7,000 people and arresting nearly 50,000.

Figure 2.3: Field practicum in the 1930s. Source: The City History Compilation Committee of Seoul.

After the March 1 movement spread across the territory, the Governor-General's administration began to relax a number of restrictions, including in education. In 1922, it elevated the National Engineering Training Center to the level of a school, meaning it provided not only training but also education. The administration also began permitting some students to seek education in Japan. These were typically students of means, especially former *yangban* families, and they sponsored themselves. Approximately 4,500 students studied in Japan in 1923. The majority of those who traveled to Japan enrolled in either academic or technical secondary schools. The proportion of study-abroad students who attended universities always remained less than 10% of the total, with roughly half of those in engineering or the sciences.

A much smaller number gained permission to travel to the U.S., primarily through support from Protestant Christian missions, and especially to see higher education. Missionaries also promoted Christianity and resisted Japanese imperialism by offering professional education through private colleges. Christian colleges such as Yonhi College (Presbyterian), Soongsil Academy

(Presbyterian), and Ewha School for Women (Methodist) offered professional education to "lay a foundation for a 'nourishing and integral' Christian life."[31] Still, those students who succeeded in completing some form of higher education rarely found work in colonial Korea adequate to their skills and knowledge.

Throughout the Japanese occupation, a significant majority of those students who accessed mid- to high-level education across Joseon were actually from Japan. In 1924, the Governor-General established an imperial university in Joseon in order to provide higher-skilled labor in medicine, law, and administration. Of the 66 students who initially enrolled in medicine at the new imperial university, 52 were Japanese, as were 51 of the 84 enrolled in law. One Korean student at the *Kyeong-seong* Higher Engineering School in 1927 reported that only 70–80 students were in attendance at the time, of which three quarters were Japanese.[32] The number of students in mid-level technical education did increase from roughly 3,000 in 1919, at 88 schools, to more than 12,000 in 1935, at 116 schools. But throughout this period, the percentage of Koreans at these schools never exceeded 25%. The newspaper *Joseon Daily Editorials* (*Joseon Ilbo*) quoted a Japanese businessman complaining in 1936 that "if we . . . want to do business in Joseon, we have to bring technicians from Japan."[33] This was largely true. In 1941, 98% of technicians working in industries across the peninsula had come from Japan.[34] During the 36 years of colonization, science and engineering graduates totaled about 400 Koreans. Only 12 acquired Ph.D.'s.[35]

The empire's plan for Joseon industry was to extend large industrial conglomerates onto the peninsula, using its human and natural resources to support minor subdivisions. This plan lasted until the Japanese invaded China in 1937. The Japanese military, which controlled the government at home, established the "Continental Military Logistics Base" on Joseon in order to take advantage of its strategic location. From then until the beginning of the Pacific War with the U.S., the support roles played by Joseon industries were effectively military in content, with Japanese personnel in leadership positions.

A prime example was the textile industry. In 1938, the year Joseon achieved its highest output in textiles, a large factory owned by Joseon Textiles employed more than 500 men and nearly 3,000 women in technical support positions. The factory employed seven male engineers, all of whom were Japanese. Only one manufacturer employed any Korean engineers, Kyeong-seong

---

[31] Shin, "Singminji Joseonui Godeung Gyo-Yuk Chegyewa Mun Sa Cheorui Jedohwa Geurigo Singminji Gong Gong Seong (The Higher Education System and Institutionalization of Literature, History, and Philosophy in Colonial Joseon)," 2012, 69.

[32] Lee, *Hangug Gisul Gyoyuksa* (*Korean History of Technology Education*), 1991, 254.

[33] Abe, *Joseon Tongchi-Ui Haebu* (*Analysis of the Colonial Rule of Joseon*), 1927, 96.

[34] Lee, *Hangug Gisul Gyoyuksa* (*Korean History of Technology Education*), 1991, 319.

[35] Song, *Gwahak Gilsulgwa Sahoe-Ui Jeob Jeomeul Chajaseo* (*Points of Contact among Science, Technology, and Society*), 2011, 121.

Textile, which employed 11, again all male. That company was founded and run by Koreans (see Figure 2.4).

Figure 2.4: Kyeong-seong Textile. Source: The City History Compilation Committee of Seoul.

After the Japanese repressed the independence movement, two pathways to resistance scaled up across the peninsula. One was an odd mix of Social Darwinism and a Confucian commitment to elevated spirit. As had been the case in Japan,[36] the Darwinist image of "survival of the fittest" appealed to a variety of elites across Joseon. The formerly dominant *yangban*/Confucian practice of disciplining the individual spirit or soul still took priority over engaging the material world, but now with the addition of new elements. With life chances dominated by an outside military force, the focus on spirit and soul honed in on ways to hang on to the Korean language and to forge a specifically Korean history. Korea was at risk of disappearance, but it would survive if its people preserved and nurtured a distinctive set of beliefs and convictions. The stories of such efforts are legion and powerful.[37]

A second pathway was actually to scale up an embrace of science. In 1924, Kim Yong-kwan (1897–1967) established the Invention Society. After graduating from a mid-level technical school,

[36] Kinmonth, *The Self-Made Man in Meiji Japanese Thought*, 1981.

[37] Chun and Ko, "Geundae Hangugui Sahoe-Jinhwaron Doibe Boineun Jeongchijeok Insik Gujo (On the Significance of Social Darwinism in the Late Joseon Dynasty)," 2011, 40; Woo, "*Sahoe Jinhwa-Ronkwa Minjokjuui (Social Evolution and Korean Nationalism),*" 2008, 161–163.

Kim had studied ceramics for a year in Japan. Under his leadership, the Invention Society published the science magazine *Science Joseon* (과학조선). A decade later, in 1934, it sponsored Science Day to commemorate the birth of Charles Darwin. The day's slogan, "The winner in science is the winner of all," illustrated the Society's goal of keeping Joseon citizens informed of developments in science in order to keep alive the dream of liberation. And since it could claim it was simply distributing scientific information, not establishing a school nor advancing an explicitly liberationist vision, the magazine persisted. The Invention Society made itself the hub of a larger social movement to popularize science.[38]

When the Pacific War (1941–1945) dramatically increased demands for Japanese technical personnel, they began returning to the archipelago. Koreans who had acquired know-how while working for long periods at low-skilled positions began to flow into the vacancies left behind.[39] Opportunities for technical education also expanded. Between 1936 and 1943, the number of commerce schools increased from 14 to 22, agricultural schools from 30 to 54, and engineering/industry schools from one (*Kyeong-seong* Higher Engineering School) to ten. Still, these numbers were rather trivial across a territory whose population exceeded 20 million. Also, the quality of instruction in engineering was far below the levels offered in Japan. In 1942, only 15 students graduated in textiles, 17 each in architecture and applied chemistry, 18 in civil engineering, 30 each in mechanics and electrical technology, and 130 in mining. The total for all of Joseon was 307.

Toward the end of the colonial period, Dr. Yamaga Shinji (1887–1954), the imperial university's last president who had worked to introduce education in science and engineering, summarized both the recent growth and yet diminutive size of Joseon industry. "From 1929 to 1939, the chemical industry grew 460%," he reported, followed by the metal industry, which experienced a "100% increase during the two years following 1937." Things looked very differently, however, when compared to the empire as a whole. "[I]f we set Japan's total industrial output at 100," he said, "Joseon is a mere 4.5%."[40]

During the period of colonial rule, the Japanese had especially resisted higher education in engineering. They made small concessions toward the end. In 1941, when the military redirected its attention away from China and toward Southeast Asia and the Pacific, the Governor-General established a science and engineering department at the imperial university—fully 17 years after its founding. Once again, we must not be deceived. It produced only 36 graduates over its first three years.

[38] Park, *Hankugsa-Edo Kwahak-Eun Inneunga? (Does Science Exist in Korean History?)*, 2004, 270.

[39] Kim, *Hanguk Geundae Gwahak-Gisul-Illyeogui Chul-Hyeon (The Emergence of Korean Science-Technology Personnel)*, 2005, 449.

[40] Lee, *Hangug Gisul Gyoyuksa (Korean History of Technology Education)*, 1991, 304.

These graduates would play important roles across the peninsula after liberation in 1945. But until then, the colonial government had no place for Koreans who had completed higher technical education. They had to make their way as best they could.

NO PLACE FOR ENGINEERS IN AN AGRARIAN VISION: 1945–1948

Liberation from the Japanese empire on August 15, 1945, did not generate a tidal wave of interest in engineering across the new country. Not even a small wave. Initially, there was not even a country. Then there was a country relatively lacking in places for engineers.

After the Japanese empire surrendered to the Americans in 1945, many Koreans expected independence to follow quickly. Across the peninsula, "peoples' committees" formed with the goal of creating a new, more egalitarian country. Indeed, within weeks leaders in Seoul announced the formation of the People's Republic of Korea.[41]

Yet the terms of surrender partitioned the Korean peninsula along the thirty-eighth parallel, just north of Seoul, into zones occupied by military forces from the U.S. and the Soviet Union. The Soviet Union had declared war on Japan shortly before it surrendered. To American military and diplomatic officials focused on Japan, Korea was a "side-show," an "unexpected, largely unwanted responsibility."[42] Officially, the U.S. and Soviet governments were collaborating under United Nations leadership to prepare Korea for independence. But during an occupation that lasted three years, the U.S. Military Government in Korea, USAMGIK, never sought to develop a deep understanding of Korean politics, history, or daily life. Relying entirely on local translators to gather information and communicate policies and decisions, USAMGIK never seriously engaged in country-building.

The U.S. and Soviet Union would never come close to agreeing on a common pathway for their two zones. U.S. officials were obsessed with the threat of communism, and China quickly became a model of what could happen in Korea. When the Pacific War ended, civil war in China resumed between the nationalists under Chang Kai-shek and the Communist Party under the agrarian Marxist Mao Zedong. The war raged for four years until the communists finally drove the nationalists to Taiwan.

For USAMGIK, fighting communism in the American-controlled zone took two forms. One was explicit repression. The occupying military refused to recognize a new people's republic, "dash[ing] hopes of a unified Korea."[43] It mobilized the centralized police force created by the Japanese to suppress waves of protests and strikes by, especially, students and peasants. And as the U.S. government would do so many times throughout the world, it gave primacy to the litmus test of anti-communism in its assessments of friends and enemies. The result was typically support for conservative elites who declared themselves to be anti-communist. To elites in the U.S.-controlled

[41] Seth, *Education Fever*, 2002, 42.

[42] Seth, *Education Fever*, 2002, 42.

[43] Lie, *Han Unbound*, 1998, 7.

zone, the nearby war in China made it clear that the egalitarian politics and policies championed in communist philosophies directly threatened their authority. Such leaders were generally more than willing to embrace a strict anti-communism and ally themselves with the U.S. military. By 1947, USAMGIK had closed down more than 57 of the zone's secondary schools and mid-level colleges in order to suppress dissent.

The second approach USAMGIK took to fighting communism was nearly the inverse of the first. It involved promoting democracy through decentralized education. The military government expressed strong interest in replacing the militarist educational practices the Japanese had left behind. It also promoted educational practices and policies that championed broad participation, equal opportunity, self-reliance, individual responsibility, and the nurturing of Korean identity.[44]

The resulting ambivalence and, hence, ambiguity in U.S. policies became obvious to all when the military government created Seoul National University in 1946. The image of education that the American occupiers brought with them necessarily included higher education to produce intellectual leaders for a new country. The plan included a graduate school to educate scholar-leaders. The military government created Seoul National University by forcing together ten different institutions with ten distinct local identities. Beginning with the former *Kyeongseong* Imperial University, it added the *Kyeongseong* colleges of law, economics, medicine, dentistry, mining, and industrial engineering, as well as the men's normal school (teachers' college), women's normal school, and Suwon agricultural college. It then grouped these into nine colleges, including a college of liberal arts and science and a college of engineering. The U.S. military claimed that the financial stability provided by this consolidation would offer students a wider range of choices and education officials a more economical, efficient use of instructional and building resources.

To left-leaning students and faculty dreaming of a more egalitarian country, the image of a national university signaled something else entirely—a continuation of centralized control over education. Seoul National University would not be a people's school. It would not extend the relative autonomy of its formerly separate units. It would become one large organization directly under the supervision, or thumb, of a conservative ministry of education. On top of it all, the American military appointed an outsider, an American, as the university's first president. Fierce opposition to the move lasted a year and a half, until USAMGIK finally fired more than 300 faculty. Nearly 5,000 students left the school.[45]

The university's two initial deans (see Figure 2.5) found themselves caught in an intense postwar conflict over the future of Korea between 1945 and 1950 that would ultimately claim 100,000 lives.[46] When Dr. Ree Tai-kyue (1902–1992) agreed to return from Japan and lead the college of

---

[44] Seth, *Education Fever*, 2002, 36.

[45] Committee, *Seoul Daehaggyo 60 Nyeonsa* (*A 60-Year History of Seoul National University*), 2006, 20–22.

[46] Lie, *Han Unbound*, 1998, 8; Kim, "Nambugui Du Gwahagja Ree Tai-Kyue Wa Li Seung-Ki (Two Scientists from South and North Korea: Ree Tai-Kyue and Li Seung-Ki)," 2008, 19.

arts and science, he announced, "I am the son of my country. Sacrificing my life as remuneration is my duty."[47] He had been one of those few who had earlier gained admission to Kyoto Imperial University in Japan. Ultimately earning a Ph.D., he gained recognition for introducing quantum chemistry to Japanese science. Swept into the internal dispute over the character and contents of Seoul National University, Ree Tai-kyue became so disillusioned that he left for the U.S. in 1948. At that time, scientists in the U.S. were championing freedom for scientific investigation. The U.S. National Science Foundation would be established shortly thereafter. Perhaps Dr. Ree found this emphasis appealing.

Meanwhile, Dr. Li Seung-ki (1905–1996) returned from the Dakaski research institute in Japan to become dean of the college of engineering. He had been the first Korean to earn a Ph.D. in engineering, also at Kyoto. He would later become known as the inventor of the synthetic fiber vinalon. On his appointment, Li Seung-ki expressed pride in "offer[ing] my people welfare, by freely unfolding my scientific visions in Korea, which is something I could not do freely during the Japanese rule."[48] In his case, a combination of political tension and the closure of a factory that had served as his research site drove him north.

Dr. Li Seung-ki was actually joining a flood of engineers and scientists flowing north. Prospects for industrial development in the north were certainly part of the attraction. Most Japanese industry had been in the north, and it held the bulk of the peninsula's mineral resources. The Soviet Union's political ideology emphasized liberation for subordinated masses through large-scale industrial development and leadership by a technocratic Communist Party. A significant majority of its governing body, the Politburo, held degrees in engineering.[49] The new Kim Ilsung University in the north supported the Soviet Union's plan for rapid industrialization by privileging its colleges of engineering and medicine, while also rapidly expanding majors in the sciences. Attracted by the seeming opportunity to become leaders of a new Korea, the engineers and scientists who went north produced a brain drain that seriously hampered developments in the south.[50]

In the south, savvy technicians who had long labored in low-level positions found themselves with opportunities to move up. Some technicians and low-level engineers became entrepreneurs, purchasing usable factories, especially in the textile industry, that the military government had appropriated after the war and put up for sale.[51] But these self-defined opportunities for techno-national formation were more the exception than the rule. The occupying U.S. force was not committed to building on the installed capacity left behind by the Japanese, and it provided little assistance

[47] Kim, *Auneu Gwahakja-Ui I-Ya-Gi* (*A Scientist's Story*), 1990, 97.

[48] Li, *Gyeore-Ui Kkum Gwahage Sireo* (*Science, the Road to the People's Future*), 1990, 33.

[49] Graham, *The Ghost of the Executed Engineer*, 1993.

[50] Kim, "Kim Ilsung Jonghab Daehagui Changnip Gwa Bunhwa (The Foundation and Division of Kim Ilsung University in North Korea)," 2000, 216. Unfortunately, we are not able to provide reliable numbers here.

[51] Kim and Seo, *Hanguk Myeonbangjik Gong-Eopui Baljeon* (*Development of the Korean Cotton Textile Industry*), 2006.

to those Koreans who were.[52] In the American diplomatic vision, Korea was an agrarian country without the intellectual or natural resources to support industrial development. Administrative policies envisioning an agrarian economy and focusing on anti-communism effectively supported the landlord class. U.S. assistance came primarily in the form of consumer goods.

Figure 2.5: Dr. Ree Tai-kyue (1902–1992) and Dr. Li Seung-ki (1905–1996). Source: Yuksa-Bipyeongsa Editorial Board 2008: 176.

By the end of the occupation in 1948, manufacturing production in the south was only 15% of that achieved under Japanese rule in 1939.[53] In the south, the six-year vocational schools (middle school plus high school) had only 26,000 students in 22 industrial divisions, and the terminal vocational middle schools had only 1,500 students in 5 industrial divisions.[54] Seoul National University had few books and inadequate facilities. No administrative office in the new Republic of Korea bore direct responsibility for education in engineering or training in technical skills.

REBUILDING AGAIN WITHOUT ENGINEERS: 1948–1960

During the next two years, the new national assembly of South Korea began seeking ways of expanding both education and industry, but funds were scarce for supporting any initiatives. These efforts ended abruptly when military forces from North Korea invaded the south in 1950.

[52] Seo, "Haebang Jeonhu Daehyumo Myeonbangjik Gongjang-Ui Gogeupgisulja (A Study of Highly Qualified Engineers at Large-Scale Cotton Textile Factories before and after Liberation)," 2006, 82.

[53] Kim, *Hanguk-Ui Gyo-Uggwa Gyeongjebaljeon: 1945–1975* (*Korea Education and Economic Development: 1945–1975*), 1980, 78.

[54] Lee, *Hangug Gisul Gyoyuksa* (*Korean History of Technology Education*), 1991, 323.

The North Korean army successfully gained control of all but two cities in the south. In its sweep across the country, it systematically destroyed all the factories and other industrial facilities. The U.S. entered the war in 1951 to stop what it saw as the dangerous spread of communism. By 1953, three million lives had been lost. Both South Korea and North Korea would find themselves having to start over again building the means to support their populations, let alone expanding industry. They would also find themselves at the center of a Cold War conflict between two superpowers.

Before the war ended, South Korea began implementing a land reform policy that the national assembly had approved prior to the war. Capping ownership of land at 7.5 acres had dramatic effects. The policy effectively produced in a surprisingly short time a "sea of small landholders."[55] In addition to destroying factories, the North Korean army had liberated land for peasants. Always concerned first about communism, the U.S. military "goaded the… government into pursuing land reform in the countryside in 1952, when the war was stalemated at the 38th parallel." The initiative helped build public support for the war, helping to avert defeat.[56] Toward the end of the earlier Pacific War, in 1944, the richest 3% of farming households had owned 64% of the arable land. By 1956, the top 6% owned only 18%.[57]

The broader effects of land reform were equally remarkable. In particular, ownership of land lost status as a primary indicator of rank and standing. The long-eroding *yangban* educational process, which had educated landed elites and channeled them into positions of political power, formally ended. During the 1950s, a new flow of elites into higher education and then business first supplemented and then replaced the ownership of land with the ownership of small businesses. Higher education and business became the new "royal road to prestige across the Republic of Korea."[58]

Owing in significant part to the long Japanese suppression of academic education, after the Pacific War ended the number of schools, teachers, and students exploded across the country. In 1945, the percentage of school attendance by children had been 64%. By 1954, it was 80%, and by

---

[55] Lie, *Han Unbound*, 1998, 15.

[56] Lie, *Han Unbound*, 1998, 11.

[57] Cho, "Post-1945 Land Reforms and Their Consequences in South Korea," 1964, 84, quoted in Lie, *Han Unbound*, 1998, 11–12.

[58] Lie, *Han Unbound*, 1998, 16. Lie argues convincingly that land reform grounded economic expansion during the 1960s, distinguishing Korea from Latin American countries seeking economic expansion during the same period. For further comparisons of economic development in Korea and Latin America during this period, see Frieden, "Third World Indebted Industrialization," 1981; Hughes, "Why Have East Asian Countries Led Economic Development?," 1995.

1960, over 95%. University attendance totaled approximately 8,000 in 1945. During the 1950s, the annual rate of growth in university attendance was nearly 15%, reaching 100,000 in 1960.[59]

The pre-war national assembly had taken an initial step to expand vocational education by developing schools at the secondary level. It had resisted U.S. pressure to elevate the status of technical education, however. The U.S. government wanted it to establish a single-track system for all students in junior high and high school. Instead, the post-war assembly basically restored the Japanese dual system, which had sharply separated academic from vocational tracks. The postwar explosion of desire for education aimed at academic education that would facilitate "spiritual" development in new ways through the humanities and social sciences. Vocational education still carried a stigma.

Beginning in 1951, the law provided for six years of common elementary school preparation followed by three years of middle school and three years of high school. Those on an academic track prepared for exams leading to admission to four-year universities. The vocational side had more options, including terminal middle schools, linked middle and high schools, and, ultimately, two-year junior colleges for those who had completed vocational secondary education.[60] Those schools that admitted girls focused on preparing them for the lowest-status factory work.

After the war, the U.S. government contracted with the University of Minnesota to help (re-) build Seoul National University, including support for medicine, agriculture, and engineering.[61] The $10 million in aid funds allocated to the "Minnesota Project" brought 218 professors to Minnesota for advanced education. This funding amounted to 78% of all aid for higher education. Fifteen participants acquired Ph.D.'s.[62] Yet despite this targeted effort, industrial education never became a priority for the aid agencies. In the midst of rapidly expanding interest in academic education, technical schools became mere "havens for students who failed to get into academic high schools or were poor."[63]

The U.S. did provide funds for training the South Korean military. And at a time when both the U.S. and Soviet Union were proliferating nuclear capabilities among their allies, the U.S. provided support for a Korean Office of Atomic Energy in 1959 (see Figure 2.6). Both the military and the atomic energy office required highly trained technical officials. Yet because the country lacked the necessary educational resources, the government encouraged students to look elsewhere. De-

[59] Mason and Kim, *Hanguk Kyeongje Sahoe-Ui Goendaehwa: Hanguk Kyeongje-Ui Geundaehwa Kwajeong Yeongu* (*The Economic and Social Modernization of the Republic of Korea*), 1981, 357; Chung, "5.16 Kudeta Ihu Jishiginui Bunhwawa Jaepeon (Differentiation and Reorganization of Korean Intellectuals after the May 16 Coup)," 2004, 160.

[60] Ministry of Education, *Gyo-Yuk 50 Nyeonsa* (*Fifty-Year History of Education*), 1998, 61.

[61] Dodge, "A History of U.S. Assistance to Korean Education, 1953–1966," 1971, 158–180.

[62] Jo et al., "*Hangug-Ui Gwahag-Gisul Inryeog Jeongchaeg* (*Review of Science and Technology Human Resource Policies in Korea*)," 2002, 91.

[63] Lee, *Hangug Gisul Gyoyuksa* (*Korean History of Technology Education*), 1991, 331.

mand for study abroad, mostly to the U.S., increased from roughly 1,000 students per year to 5,000 students per year, with many seeking education in engineering and the sciences. Thus, during the years after the war with North Korea, some Koreans did pursue interests in engineering and could find appropriate work to transform those interests into careers. We can also say that the Korean government initiated a kind of techno-national formation, for the purposes of national security. It is just that such formation took place, for the most part, outside of Korea.

Figure 2.6: Interior facilities of the Office of Atomic Energy (1960). Source: National Archives of Korea.

Despite these new opportunities, interests in engineering and opportunities for engineers remained exceptional, small in scale. After the Korean War, the U.S. and United Nations poured aid into the country, mostly in the form of consumer products and support for light industries to reduce imports of consumer products (see Figures 2.7 and 2.8).

Figure 2.7: President Rhee Syng-man at the ceremony for delivering relief supplies, 1953. Source: National Archives of Korea.

Figure 2.8: Ceremony to deliver relief supplies from the United Nations, 1959. Source: National Archives of Korea.

Initiatives in industry did grow rapidly under the Rhee Syng-man government (1948–1960), but these called for low-level technical capabilities that were already widespread (see Figure 2.9).

Small companies such as Samsung and Hyundai were founded by technical entrepreneurs who had trained at manufacturing sites. The country's dominant industries at the time produced wheat, raw cotton, sugar, and other agricultural products. All told, these industries had little need for technical practitioners with education beyond the secondary level. The 1950s saw no national commitment from the Republic of Korea to engineering education and work for engineers.

Figure 2.9: Sibal, the first assembled car in Korea, by the Kukjecharyang Company, 1957. Source: National Archives of Korea.

A MATTER OF INDIVIDUAL INTEREST AND AMBITION

During the 85-year period from 1875 to 1960, the Korean peninsula hosted five different Koreas. None championed the making of engineers.

The 500-year Joseon dynasty gave way to the Great Korean Empire in 1897 in an ill-fated effort to hold off the Japanese. The Joseon dynasty had long privileged male scholars. Unlike Japan, but like China, its literary scholar-officials were not also warriors. The landholding class, *yangban*, held elite status. Its male members gained and held privilege by memorizing and reciting classic Chinese texts. Work with one's hands put one far from the practice of understanding connections between humans and nature as a whole. We can speculate that the Great Korean Empire may have ultimately ventured into higher engineering education for men, given that the Japanese had done so successfully. But the Korean empire was gone before it had the chance.

Victories in war with China and Russia enabled the Japanese empire to make Korea a protectorate in 1905. Western countries did not object when Japan annexed the Korean peninsula in 1910 and made Joseon a subordinate part of the expanding Japanese umbrella. Believing Koreans to be racially inferior and fearing that education would foster Korean identity, the occupying Governor-General's office limited education to producing low-level support for Japanese industry and other projects. One could find engineers with advanced training across the Joseon province of the Japanese empire, but they were mainly Japanese engineers. The increasingly militant Japanese government did not succeed in eliminating the Korean language nor prevent attempts to create a separate Korean history. Yet the popular science movement that also formed did not scale up into widespread interest in producing or becoming advanced technical practitioners. When the Japanese engineers began to leave late in the Pacific War, only low-level technicians were there to replace them.

After the Pacific war, dreams of independence were dashed by a superpower agreement between the U.S. and Soviet Union to divide the peninsula until they could agree on how to forge an independent Korea. They failed. Each fearful of the other, both actually wanted some kind of continued dependency. The American presence did not turn into an emphasis on technical education, unlike in the U.S. itself. American military and diplomats saw Korea as an agrarian country, similar to China. They focused only on education to generate resistance to communism, not to produce industrial development.

Formal separation of the Republic of Korea from the People's Republic of Korea began in 1948, although in 1950 North Korea nearly gained complete control of the South. The U.S intervention re-established the boundary at the 38th parallel, producing a tense standoff that continued into the 21st century. The Republic of Korea officially regained its independence, yet only by accepting a permanent U.S. military force numbering in the tens of thousands stationed just outside of Seoul. During the balance of the decade, the South Korean government began to seriously champion self-reliance and South Koreans dramatically expanded their demand for educational opportunities. But with the government's industrial interests limited to light industry, policies to enhance industrial self-reliance and educational achievement did not highlight engineering.

Throughout the 85 years, many individuals had successfully sought education in engineering and even found work as engineers. Some had gone to Japan, some to the U.S. An engineering school was established under Japanese rule. Seoul National University gained a college of engineering. Yet despite the myriad of fascinating trajectories that involved higher education in engineering, the work of becoming an engineer prior to 1960 was by and large the product of individual visions and individual ambitions.

The five Koreas of 1875–1960 did not embrace them, which is to say that no techno-national image of engineers and engineering successfully scaled up that linked them to dominant images and trajectories of the territory as a whole. None of the five Koreas looked to engineers to lead or even occupy positions with stable identities. Engineers were simply faces in Korean crowds struggling to make their way through challenging and rapidly changing circumstances.

REFERENCES

Abe, Kaoru. *Joseon Tongchi-Ui Haebu* (*Analysis of the Colonial Rule of Joseon*). Kyungsung: Minjung-sironsa, 1927. 35

Board, Yuksa-Bipyeong Editorial. *Namgwa Bugeul Mandeun Raibeol* (*Rivals Who Constructed South Korea and North Korea*). Seoul: Yuksa-Bipyeong-sa, 2008.

Boulesteix, Frederic. *Chakan Migae-in Dongyangui Hyeonja* (*Good Savage and the Oriental Wise*). Seoul: Cheongnyeonsa, 2001. 26

Cha, Mi-hee. "Joseon Sidae Mungwa Yeongu-Ui Donghyang-Gwa Jeonmang (Trends and Prospects About Mungwa in Joseon Period Research)." *History Education Review* 49, no. 0 (2012): 287-320. 27

Cho, Jae Hong. "Post-1945 Land Reforms and Their Consequences in South Korea." Unpublished Ph.D. diss., Indiana University, 1964. 42

Chun, Ja-hyeon and Hi-tak Ko. "Geundae Hangugui Sahoe-Jinhwaron Doibe Boineun Jeongchi-jeok Insik Gujo (on the Significance of Social Darwinism in the Late Joseon Dynasty)." *Daehan Political Science Review* 18, no. 3 (2011): 27-48. 36

Chung, Yong-wook. "5.16 Kudeta Ihu Jishiginui Bunhwawa Jaepeon (Differentiation and Reorganization of Korean Intellectuals after the May 16 Coup)." In *1960 Nyeondae Hangugui Geundaehwawa Jishigin* (*Korean Modernization and Intellectuals in the 1960s*), edited by Chung, Yong-wook, Seoul: Sunin, 2004. 43

Committee, SNU History Compilation. *Seoul Daehaggyo 60 Nyeonsa* (*A 60-Year History of Seoul National University*). Seoul: SNU Press, 2006. 39

Dodge, Herbert W. "A History of U.S. Assistance to Korean Education, 1953-1966." George Washington University, 1971. 43

Downey, Gary Lee and Masanori Wada. "Avoiding Inferiority: Global Engineering Education across Japan." Paper presented at the American Society for Engineering Education Conference & Exposition, Vancouver, 2011. 28

Elman, Benjamin A. *A Cultural History of Modern Science in China*. Cambridge, Massachusetts: Harvard University Press, 2006. DOI: 10.1002/jhbs.20412. 25

Frieden, Jeff. "Third World Indebted Industrialization: International Finance and State Capitalism in Mexico, Brazil, Algeria, and South Korea." *International Organization* 35, no. 3 (1981): 407-431. DOI: 10.1017/S0020818300032525. 42

Graham, Loren R. *The Ghost of the Executed Engineer: Technology and the Fall of the Soviet Union*. Cambridge, Mass.: Harvard University Press, 1993. 40

Hughes, H. "Why Have East Asian Countries Led Economic Development?" *Economic Record* 71, no. 212 (1995): 88-104. DOI: 10.1111/j.1475-4932.1995.tb01874.x. 42

Hyeon, Kwang-ho. "1880 Nyeondae Huban - 1890 Nyeondae Jeonbangi Joseonui Peurangseu Insik (French Understanding of Joseon between the Latter 1880s and the Early 1890s)." *Humanities Studies* 44, no. 0 (2012): 445-476. 26

Jang, Kyung-nam. "Joseonhugi Yeonhaeng-Rokui Cheonjudang Gyeonmunroggwa Seohak Insik (an Envoy's Observational Record of the Catholic Church in Late Joseon and a Discussion of Western Learning)." *The Studies of Korean Literature* 26, no. (2009): 77-117. 25

Jo, Hwanghee, EK Lee, CG Lee and SW Kim, "Hangugui Gwahag-Gisul Illyeok Jeongchaeg (Review of Science and Technology Human Resource Policies in Korea)," Science and Technology Policy Institue, 2002. 43

Kim, Geun-bae. "Kim Ilsung Jonghab Daehagui Changnip Gwa Bunhwa: Gwahak Gisulgye Hagbu-Reul Jungsimeuro (The Foundation and Division of Kim Ilsung University in North Korea: Focusing on the Technical Disciplines)." *The Korean Journal for the History of Science Society* 22, no. 2 (2000): 192-216. 40

Kim, Geun-bae. *Hanguk Geundae Gwahak-Gisul-Illyeogui Chul-Hyeon (The Emergence of Korean Science and Technology Personnel)*. Seoul: Moonji, 2005. 32, 33, 37

Kim, Geun-bae. "Nambugui Du Gwahagja Ree Tai-Kyue Wa Li Seung-Ki (Two Scientists from South and North Korea: Ree Tai-Kyue and Li Seung-Ki)." *Yoksa Bipyeong*, no. (2008): 16-40. 39

Kim, Han-jong. "Joseon Chongdogbu-Ui Gyoyuk Jeong Chaeggwa Gyogwaseo Balhaeng (Educational Policy and the Publication of Textbooks by the Joseon Government-General)." *Studies on History Education*, no. 9 (2009): 295-329. 31

Kim, Jong-gyu and Moon-seok Seo. *Hanguk Myeonbangjik Gong-Eopui Baljeon (Development of the Korean Cotton Textile Industry)*. Seoul: The National History Compilation Committee, 2006. 40

Kim, Sang-bae. "Geundae Hangugui Gisul Gae-Nyeom (The Concept of Technology in Modern Korea)." In *Geundae Hangugui Sahoegwahak Gaenyeom Hyeongseongsa (Concept Formation in Social Science of Modern Korea)*, edited by Ha, Yeongseon, Sohn Yeol, 307-341. Changbi, 2012. 23, 25, 30, 31

Kim, Yeon-hee. "Daehanjeguggi Saeroun Gisulgwanwon Jibdanui Hyeongseonggwa Haeche (The Establishment, Development, and Disbanding of Telegraph Technicians in Imperial Korea)." *Korean Historical Studies*, no. 140 (2008): 183-220. 29

Kim, Yeong-bong. *Hanguk-Ui Gyo-Uggwa Gyeongjebaljeon: 1945-1975* (*Korea Education and Economic Development: 1945–1975*). Seoul: Korea Development Institute, 1980. 41

Kim, Yeong-ho. "Jeong Dasanui Gwahak-Gisul Sasang (Jeong Dasan's Thought on Science and Technology)." *Asian Studies* 19, no. 1 (1989): 277-300. 23

Kim, Yong-deok. *Auneu Gwahakja-Ui I-Ya-Gi* (*A Scientist's Story*). Seoul: Donga, 1990. 40

Kinmonth, Earl H. *The Self-Made Man in Meiji Japanese Thought : From Samurai to Salary Man.* Berkeley, Calif.: University of California Press, 1981. DOI:10.2307/2598937. 36

Lee, Won-ho. *Hangug Gisul Gyoyuksa* (*Korean History of Technology Education*). Seoul: Moonumsa, 1991. 32, 33, 35, 37, 41, 43

Li, Seung-gi. *Gyeore-Ui Kkum Gwahage Sireo* (*Science, the Road to the People's Future*) Seoul: Daedong, 1990. 40

Lie, John. *Han Unbound: The Political Economy of South Korea.* Stanford, CA: Stanford University Press, 1998. 33, 38, 39, 42

Mason, Edward Sagendorph and Man-je Kim. *Hanguk Kyeongje Sahoe-Ui Goendaehwa: Hanguk Kyeongje-Ui Geundaehwa Kwajeong Yeongu* (*The Economic and Social Modernization of the Republic of Korea*) Seoul: Korea Development Institute, 1981. 43

Ministry of Education. *Gyo-Yuk 50 Nyeonsa: 1948–1998* (*Fifty-Year History of Education: 1948–1998*). Seoul: MOE, 1998. 43

Ministry of Education, Science and Technology, „ "Gwahaggisul 40 Nyeonsa (Forty-Year History of Korean Science and Technology)," Seoul: Ministry of Education, Science and Technology, 2008. 32

Park, Chon-hong. *Maehogui Jilju Geudae-Ui Hoengdan* (*Railway, Fashioning Run, Crossing Modern Time*). Seoul: Sancheorum, 2003. 28

Park, Seong-rae. *Hangugsa-Edo Gwahageun Inneunga?* (*Does Science Exist in Korean History?*). Seoul: Kyobobook, 2004. 26, 32, 37

Park, Seong-rae. *Science and Technology in Korean History: Excursions, Innovations, and Issues.* Fremont, Calif.: Jain Pub. Co., 2005. 26

Seo, Moon-seok. "Haebang Jeonhu Daehyumo Myeonbangjik Gongjang-Ui Gogeupgisulja (A Study of Highly-Qualified Engineers at Large-Scale Cotton Textile Factories before and after Liberation) " *Asian Studies* 40, no. 0 (2006): 65-85. 41

Seo, Moon-seok. "Geundaejeok Myeonbangjik Gongjang-Ui Deungjang-Gwa Gisul-Illyeok Yangseongjedo-Ui Hyeongsung (Emergence of the Large-Scale Cotton Textile Factory and Formation of a Training System for Engineers).” *Asian Studies* 50, no. 0 (2011): 119-140. 33

Seong, Jwa-kyeong. *Gisul-Ui Yihae Geurigo Hangug-Ui Gisul (The Understanding of Technology, and Korean Technology)*. Incheon: Inha Univeristy Press, 1986. 25, 30

Seth, Michael J. *Education Fever: Society, Politics, and the Pursuit of Schooling in South Korea*. Honolulu, HI: University of Hawai'i Press, 2002. 38, 39

Shin, Ju-back. “Singminji Joseonui Godeung Gyo-Yuk Chegyewa Mun Sa Cheorui Jedohwa Geurigo Singminji Gong Gong Seong (The Higher Education System and Institutionalization of Literature, History, and Philosophy in Colonial Joseon).” *The Korean Journal of History of Education* 34, no. 4 (2012): 57-81. 35

Song, Eung-seong. “Cheon-Gong-Gae-Mul (Heaven-Artisan-Open-Matter).” <manuscript>, 2009. 25

Song, Sung-soo. “Jeong Yak-Yong's Thoughts on Technology.” *Journal of the Korean History of Science Society* 16, no. 2 (1994): 261-276. 24

Song, Sung-soo. *Gwahak Gilsulgwa Sahoe-Ui Jeob Jeomeul Chajaseo (Points of Contact among Science, Technology, and Society)*. Seoul: Hanul, 2011. 35

Wada, Masanori. “Engineering Education and the Spirit of Samurai at the Imperial College of Engineering in Tokyo, 1871-1886.” Virginia Tech, 2008. 28

Woo, Nam-sook. “Sahoe Jinhwa-Ronkwa Minjokjuui (Social Evolution and Korean Nationalism: Concerning on the Park Eun Sik).” *The Review of Korean and Asian Political Thoughts* 7, no. 2 (2008): 139-167. 36

Yi, Sang-il. “Kim Yun-Shikui Gaehwa-Jagang-Ron-Kwa Yeongsunsa Sahaeng (Kim Yun-Sik's Enlightenment Thought and Yeongsunsa).” *Korea Culture Studies* 11, no. 0 (2006): 93-115. 30, 31

CHAPTER 3

Technical Workers for Light Industry: 1961–1970

A passage in the famous book pictured a young girl sitting in the second-class compartment of a train, her hands holding a book of French poetry. "Your white hands," the author asserted, "I abhor." Clean hands, he continued, "have been responsible for our present misery." Smooth hands were the hands of the privileged class. That class, and privilege consciousness in general, had to be replaced. "We must work," the author continued, because "one cannot survive with clean hands."[1]

The author was Park Chung-hee (1917–1979) and the book his 1963 volume *The Country, The Revolution, and I*. A military man, Major General Park (Figure 3.1) had taken control of the South Korean government in 1961 through a coup d'etat. Yet his forces encountered remarkably little opposition. And although he promised to stay out of partisan politics, he arrived with a clear, distinctive program for what he called the "reconstruction" of Korea. That program gave both visibility and prestige to technical workers for the first time, from semi-skilled technical labor to research scientist-engineers.

Figure 3.1: General Park Chung-hee (1917–1979). Source: National Archives of Korea, JoongAng Ilbo Photo Archive.

---

[1] Park, *Gugga, Hyeogmyeonggwa Na* (*The Country, the Revolution, and I*), 1970, 178–179.

In this chapter we examine the first phase of Park Chung-hee's program for producing a new Korea, focusing especially on techno-national initiatives in secondary and higher technical education. Following Park's own metaphor labeling two economies, we describe the ambitious initiatives of his government during its first decade and call attention to the considerable resistance he faced. Park Chung-hee would get his industrial development, what he called the "first economy" in his extensive writings. The first economy would consist primarily of light industries, whose technical workers, frequently women, could generally not be promoted as equivalent to Western engineers. Furthermore, neither he nor his government would successfully scale up broad popular support, let alone acceptance, for an image of technical labor as an icon of national unity and advancement. A cleavage developed between the government and the general population. Severe difficulties lay in what President Park called the "second economy."

A PROGRAM IN TWO PARTS

Park Chung-hee's vision for Korea had two related parts. He called one part the "'Economy First' principle."[2] Park's Korea would dramatically expand industrial production specifically in the private sector. Building local industries would reduce the country's reliance on U.S. aid for consumer products. They would enable Korea to begin escaping from dependence on a foreign country. Equally important, implementing the Economy First principle would terminate the cozy relationships between political leaders and business entrepreneurs that had developed during the regime of President Rhee Syng-man (1875–1965). Rhee had resigned following his re-election in 1960, amid country-wide protests led by students that he had rigged the election to preserve corrupt leadership by elites. The Economy First principle was designed to appeal to Koreans of all classes.

The second part of Park's program was to end partisan bickering and competition among what he called political "cliques." Koreans, according to Park, needed to unite. In particular, they needed to collectively embrace the "common ideology of anti-communism."[3] The government in the north, the People's Republic of Korea, championed communism. Communist-led governments offered Marxist visions of proletarian or peasant rule replacing illegitimate exploitation by bourgeois capitalists and landholders. The Soviet Union pursued a Marxist/Leninist/Stalinist pathway and China a Marxist/Maoist pathway. All communist countries had officially committed to eliminating hierarchical class barriers through Communist Party leadership and technological development.

At the time, communist images were immensely appealing to peasants across the Korean peninsula and elsewhere in East Asia. In virtually every case, peasants had never experienced anything approaching equality with landlords, literati, and other elites. Across South Korea, rural *yangban* aristocrats had controlled local political and economic hierarchies. During the 1950s, North

---

[2]   Park, *Our Nation's Path*, 1970, 186.

[3]   Park, *Our Nation's Path*, 1970, 186.

Korea's recovery from the war appeared to South Koreans "to be faster, its economy more robust and self-reliant, and its claims to represent a more viable model of Korean nationhood increasingly more convincing."[4]

Seeking to distinguish itself from the communist north, the Park government offered South Koreans a different pathway for improving economic conditions, yet still designed to serve everyone, in principle. It was a plan for economic development that promised economic prosperity they could achieve freely and collectively as individuals, without governmental ownership or control. Under Park's leadership, Koreans would work to build and ultimately unify a great homogeneous country that would someday catch up to and compete with the most powerful industrialized countries of the world. In contrast with the communist north, they would do this through political democracy and direct participation in free markets. And decreasing dependence on foreign aid would increase the country's ability to resist political pressure/interference even from the U.S.[5]

Park's program was actually in line with the U.S. government's new approach to fighting communism, by fostering economic development in countries deemed at risk. The U.S. had begun encouraging what it called "developing" countries to build the industrial capacity necessary to substitute domestic production for imports and then initiate private-sector, export-led growth. Only then, in the U.S.'s official diplomatic view, could democracy flourish and the evils of communism be resisted.[6]

Under pressure from the U.S. government to democratize, General Park formally ended military rule in 1963. It might be more accurate, however, to say that he made military rule informal. Park resigned his commission in the army but then immediately made himself a candidate for the presidency. He won, and then appointed senior military advisers to be the first senior advisers in his civilian government (Figure 3.2).

President Park purged his administration of elite scholar-officials trained in law, philosophy, and other areas of the humanities and social sciences. He then virtually flooded it with military personnel, who disproportionately represented families with lower- or middle-class status. During the first decade of Park's rule, nearly 70% of senior administrators had risen through military schools and training (Table 3.1). Note that zero came from the natural sciences or engineering. The educational machinery for producing scholar-officials from these fields was not yet in place.

4   Seth, *Education Fever*, 2002, 117–118.

5   Rodrik, "Getting Interventions Right," 1995, 61; You, *Gyeongje Seongjang Sinwha-Ui Heowa Sil* (*The Truth and Falsity of Economic Growth*)," 2011, 51; Hahn, "Administrative Capability for Economic Development: The Korean Experience," 1995, 8.

6   Park and Kim, "Park Chung-Hee Sidae-Ui Gugga-Gi-Eop Kwangae-E Daehan Jaegeomto (Rethinking State-Business Relations in Park's Era)," 2010, 138.

Figure 3.2: President Park Chung-hee (1917–1979). Source: National Archives of Korea.

Table 3.1: Fields of study for Rhee and Park administrators. Source: Han Seung-jo 1975:7[7]		
Major	% Rhee administrators (1948–1960) n = 148	% Park administrators (early years) n = 47
Law	**31.9**	10.9
Economics	10.7	13.0
Politics	9.0	2.2
Philosophy	**13.9**	0
Military	6.6	**69.5**
Natural science/ Engineering	9.0	0
Literature	4.9	0
Medicine	4.1	2.2
Others	9.0	2.2
Total	100 %	100 %

Looking back from nearly a decade later, in 1970, it was clear that domestic industrial pro-duction did indeed grow rapidly. Yet the Park government exceeded the constitutional two-term limit on the presidency and was still in power. As we will see, Park's program for democracy and economic freedom came to depend strongly on visible and vocal commitments to discipline and

7   Han, "Hanguk Jeongchi Elit-Ui Chungwon Yuhyeong (The Recruiting Pattern for Korean Political Elites)," 1975, 7.

duty, as defined by his administration. The Economy First principle became a mantra that did not necessarily link economic prosperity to expanded freedoms. The other part of the Park Chung-hee program, virulent anti-communism, came to justify explicit repression of virtually every act of resistance or opposition.

The Park regime actually lasted 18 years, until 1979. Its complex ambiguities and contradictions remain much debated by historians.[8] What stands out for our purposes is one key dimension of the Park Chung-hee rush to expand industrial production. Beginning in this first decade, it included extensive, and repeated, attempts to expand educational opportunities for technical practitioners as well as to increase their visibility and prestige. For the first time across the southern part of the Korean peninsula, a political leader sought to make (male) technical workers at all levels, from semi-skilled laborers to scientist-engineers, key techno-national agents of Korean unity and advancement.

TECHNICAL SOLDIERS FOR INDUSTRIAL DEVELOPMENT

As we explained briefly in Chapter 1, the South Korean government began categorizing technical workers in 1962, the year after President Park gained power. It is worth reflecting on the challenges facing a government wanting to reorient the priorities of an entire country. Inhabitants of the Korean peninsula had long devalued technical labor. What would it take to not only develop and mobilize the material and human resources to dramatically expand industrial productivity, but also to convince Koreans of the value of such work? Could President Park persuade Koreans to reconfigure their identities by adding industrial production to those activities they valued most highly?

After the coup, General Park made himself head of the Supreme Council for National Reconstruction. The Supreme Council in turn created the Economic Planning Board to serve as its central agency for economic planning and development. This Board faced a tricky challenge in light of Park's commitment to anti-communism, backed by the U.S. In communist countries, the Communist Party exercised centralized control over economic development. Park's predecessor, Rhee Syng-man, had shunned planning on the grounds such was a Stalinist way of thinking.[9] The Supreme Council's planning board for economic development had to figure out ways to facilitate and influence economic activity without owning or explicitly controlling it.

During its first year, the Economic Planning Board developed a comprehensive survey and classification of the Korean workforce as a whole. The classification offered an approach to seeing and interpreting a unified Korea through the economic lens of work. The key image lay in the words "as a whole." By classifying the entire Korean workforce, the Board was actually creating it in the first place.[10] The initiative effectively made the case that people did not just work jobs. They

[8] Kim, *Korea's Development under Park Chung Hee*, 2004.

[9] Lee, *Gyeong-Je Geundaehwa-Ui Sumeun Yiyagi* (*Hidden History of Economic Development*), 1999, 265.

[10] Porter, *Trust in Numbers*, 1995; Bowker and Star, *Sorting Things Out*, 1999.

worked jobs that, by virtue of their distinct functions and levels of status, belonged to and served the country as a whole.

The comprehensive survey and classification had particular importance for those involved in technical work. It elevated them as legitimate workers alongside those who occupied already valued positions, such as administrative personnel in government or business. As we described in Chapter 1 (summarized in Table 1.1), the 1962 classification introduced the three levels of craftsman (*gi-neung-gong*), technician (*gi-sul-gong*), and engineer (*gi-sul-ja*). It distinguished them by level of education and years of experience in industry. Not surprisingly, the low-level craftsmen constituted the majority in the survey. Of nearly 300,000 technical workers, more than 93% were craftspersons, with women workers at the lowest levels. Technicians accounted for less than 4% and engineers less than 3%.[11]

The Park government clearly signaled its vision for the types of leaders it wanted to champion by passing the Professional Engineers Act of 1963. While the 1962 survey had mapped the workforce, its categories did not carry specific authority on the job. Not even the highest-status men workers could appeal to them for certification or income. The 1963 act sought to provide just such authority for a new category of technical worker, the *gi-sul-sa*.

Note the replacement of *–ja* in *gi-sul-ja,* or engineer, with *–sa* in *gi-sul-sa. Sa* had long labeled scholar-officials, or male scholars with authority. Accordingly, the new *gi-sul-sa* designated not just engineers in general but engineers who possessed advanced knowledge of science and technology and their application in industry, based especially upon relevant and sufficient practical experience.[12] They were all men. It formally certified those men engineers who won the label to have authority on the job.

As the highest-level technical workers, these more scholarly engineers could count themselves worthy of techno-national status alongside graduates of higher education. With status came responsibilities. The *gi-sul-sa* should be like revered scholars: cultivated, moral, and duty-bound. A key difference was that the cultivation, moral commitments, and duties of the working scholarly engineers would all be serving the new government's mission to redefine and advance Korea through expanded industrial production.

Certifying the technical workers raised many questions. Were people interested in gaining the label? What sorts of educational institutions would produce them? What relations would they have with other technical workers? How would the country produce industrial environments in which they could find work in the first place?

[11] Economic Planning Board, *Gwa-Hag-Gi-Sul Baeg-Seo* (*Science and Technology White Paper*), 1962, 23.

[12] According to Professional Engineers Act, Act #1442, *gi-sul-sa* means a person who has highly professional knowledge and application ability based on practical experience in the field of relevant technology and is qualified as a professional engineer under the Act (Article 2). If a person who passed the examination for *gi-sul-sa,* he or she has to register with the *gi-sul-sa* register (Article 5) and the Economic Planning Board shall issue a certificate to the registered *gi-sul-sa* (Article 9).

INITIAL ATTEMPTS TO SCALE UP TECHNICAL EDUCATION

During the same year the Economic Planning Board issued the workforce classification, it began implementing a five-year plan for economic development. The plan had been developed after Rhee Syng-man resigned but was shelved in the face of the subsequent political uncertainties. Counting on significant aid from the U.S. government and through extensive borrowing from banks in Japan and Europe, the Board went to work expanding the flow of commerce by building infrastructures in transportation and communication. It funded the construction of railroads, highways, harbors, communications facilities, and projects in electric power generation.[13]

The material size and scope of these initiatives carried great symbolic significance for the Park government, to the point of spectacle. The Stalinist Soviet Union had long used images of great dams, steel mills, and other large public-owned projects to champion the broad scope of benefits brought by socialism.[14] President Park's approach, by contrast, highlighted industrialization in the private sector.

In 1962, the government recruited schoolchildren to praise the ground-breaking ceremony for what would become the massive Ulsan Industrial Area along the southeast coastline. Figure 3.3 depicts a scene from the ceremony, with the mobilized students praising the project with the words "Congratulations! Anti-Communism! Ulsan Industry!" Under the Park Chung-hee program, if you were against communism you should be in favor of private industry, and vice-versa. The link between the two was unmistakable.

Park Chung-hee loved counting as a device for motivation and control. Think about it. When you count what exists in the present, you are not only announcing what is worth counting. You are also setting yourself up to define what should exist in the future. You can specify both directions of travel into the future and expected rates of movement in those directions.[15]

Explaining why "President Park used quantitative methods," a longtime senior adviser for economic policy later wrote,

> When expressed in numbers, people easily recognized and understood the goal. It was not difficult to compare the current status at the point of planning and the expected status at completion. It was also easy to assess how far the process had come along the way.[16]

President Park used strategic counting and numerical goals to set extremely ambitious outcomes. These had the effect of fostering powerful desires among subordinates to do everything possible to attain the goals and, hence, please their leader.

---

[13] Park, *Han-Guk Yeog-Dae-Jeong-Gwon-Ui Juyo Gyeong-Je-Jeong-Chaek* (*Major Economic Policies of the Korean Government*), 2009, 111.

[14] Graham, *The Ghost of the Executed Engineer*, 1993.

[15] cf. Porter, *Trust in Numbers*, 1995.

[16] O, *Park Chung-Hee-Neun Eotteogge Gyeongjaeganggug Mandeul-Eotna* (*President Park Chung-Hee's Leadership and the Korean Industrial Revolution*), 2010, 18.

Figure 3.3: Groundbreaking ceremony for the Ulsan industrial area, February 3, 1962. Source: National Archives of Korea.

In the early years of the Park government, setting numerical goals for industrial development was possible only in those areas that could in fact develop quickly. Remember that the term *gi-sul*, meaning "technology," referred mainly to industrial machinery. Yet the country possessed little industrial machinery. Unless equipment and machinery arrived through direct aid, it had to be produced locally.

This is why the Economic Planning Board focused first on import substitution. It sought to stimulate and support a myriad of small, labor-intensive industries with low-wage workers that could replace those consumer products that had long arrived through foreign aid.[17] In addition to the work on infrastructures, the Board supported domestic industries by building oil refineries and supporting the production of cement, fertilizer, and some chemicals. As time went by, it began promoting exports in tungsten, silk, anthracite coal, squid, fish, graphite, plywood, rice, and pig bristle (for brushes).[18]

President Park made the education of technical personnel a high priority. When he first learned the details of the first five-year plan, he reportedly asked, "I wonder if we will have any

---

[17] Moon, "Park Chung-Hee Sidae Damhwamuneul Tonghae Bon Gwahakgisuljeongcheakui Jeongae (A Discourse Analysis of Science and Technology During the Park Chung-Hee Era)," 2012.

[18] Lee, "Park Chung-Hee Sidae-Ui San-Eob-Jeong-Chaek (Industrial Policy of the Park Chung-Hee Era)," 2012, 106–108.

difficulties in the technology sphere." He had noticed that the economists who had constructed the plan had given minimal consideration to technology issues, including the acquisition of equipment and supply of personnel. Since "[w]e are about to embark on a new factory construction project," he is said to have asserted, "I would like you to explain if it will be possible at our current technology level, with our current technicians." "If not," he questioned, "what kind of measures [do we need] to overcome this?"[19]

Small industry required only semi-skilled technical labor. It would be several years before the infrastructures and industries would be far enough along to provide employment for many technical workers, let alone working scholar engineers. In the early years of the Park administration, the primary focus in techno-national education became, by far, industrial training for men and women craftspersons, *gi-neung-gong*, and technicians, *gi-sul-gong*, not engineers.

Through the initial workforce classification, the Economic Planning Board had taken control over all labor statistics. These included the production of technical labor through education, which it judged to be woefully inadequate. The education component of the economic plan directly challenged the dominant image of education as aimed ultimately at the production of scholar-officials.

It sought first to shift the emphasis at the secondary level from academic education, which prepared students for the entrance examination to universities, to vocational education. In 1962, 60% of secondary students were academic, 40% vocational. The Industrial Education Promotion Law of 1963, a companion to the law certifying working scholar engineers, aimed at altering the proportions. It both funded new initiatives in vocational secondary education and established more restrictive qualifications for academic secondary education, with the goal of cutting academic enrollments in half, to 30%.[20]

Reducing the supply of students to academic higher education coupled with initiatives to limit actual enrollments in academic higher education. The Supreme Council argued that advanced education in the liberal arts was out of step with a country committed first and foremost to industrial expansion, even though preparation in the liberal arts most approximated old pathways to elite positions as scholar-administrators. The Council began a massive campaign to shift enrollments in higher education toward engineering and the applied sciences.

As a first step in gaining control over the distribution of enrollments in higher education, the educational bureaucracy introduced a system for registering degrees. Registration data generated enrollment statistics, which enabled the government to rethink the distribution of maximum enrollments. Promoting higher technical education was not just a matter of supply, however. The government also had to think through the problem of demand for graduates.

---

[19] Jeon, *Hangug-Ui Gwa-Hag-Gi-Gul Gaebal* (*Science and Technology Development in Korea*), 2010, 22.

[20] Seth, *Education Fever*, 2002, 120–121; Lee, *Sireop Gyo-Yuk* (Industrial Education), 1996, 248–249.

HIGHER-LEVEL EXPERTS FOR EXPORTS

"Export or die!" President Park Chung-hee quickly became obsessed with export growth. In order to escape from an aid economy that had kept Korea dependent upon industries in the U.S. and other countries, Park wanted to scale up industry sufficiently to be associated with, not to mention compete with, the powerful exporting nations. Also, U.S. corporations complained that Korea's efforts in import substitution were reducing their ability to market the consumer products that the Korean government was required to purchase as part of aid packages. The slogan "Export or die!" emerged as one of many the Park administration used to encourage increases in exports.

To help companies develop exportable products, the government instituted a range of novel financial policies. These included stabilizing financing for capital expansion, stabilizing currency rates, minimizing interest rates, and providing economic incentives to companies that accepted the risk of entering international markets.[21] The most important financial policy for achieving competitive advantage was "using inexpensive women [workers] with lower levels of education."[22] By 1965, for example, 260,000 women had entered the workforce. Of this total, nearly 14% had little or no schooling, 64% had completed elementary school, and nearly 8% had completed only middle school. Of the remaining 15%, nearly all high school graduates, the majority had the support of affluent families. Like many men workers, these women typically worked six full days a week, including night shifts.[23]

Because rural women typically married before the age of twenty, the Park government worked vigorously to attract them to the new light industries. It championed them alongside men as "industrial soldiers" to emphasize their "sacrifice" and "endurance." It also celebrated families that sent "dutiful daughters" to fulfill family responsibilities to serve the nation. Yet as the lowest-status workers, these women routinely found themselves disdained as "factory girls," *gong-suni*, or, worse, as invisible participants in the drive for exports.[24] Broadly accepted techno-national status was, for most, far out of reach.

Initially, the competitive advantage of Korean companies lay in the low labor costs, so they had focused on developing labor-intensive light industries for consumer products.[25] But light industry had small profit margins. Even at larger scales, it would not generate significant cash. To gain

---

[21] Park and Kim, "Park Chung-Hee Sidae-Ui Gugga-Gi-Eop Kwangae-E Daehan Jaegeomto (Rethinking State-Business Relations in Park's Era)," 2010, 138.

[22] O, *Park Chung-Hee-Neun Eotteogge Gyeongjaeganggug Mandeul-Eotna* (*President Park Chung-Hee's Leadership and the Korean Industrial Revolution*), 2010, 376–377.

[23] O, *Park Chung-Hee-Neun Eotteogge Gyeongjaeganggug Mandeul-Eotna* (*President Park Chung-Hee's Leadership and the Korean Industrial Revolution*), 2010, 376–377.

[24] Hyun-mee, "Hangugui Geundaeseonggwa Yeoseongui Nodongkwon (Modernity and Women's Labor Rights in South Korea)," 2000, 44.

[25] Moon, "Park Chung-Hee Sidae Damhwamuneul Tonghae Bon Gwahakgisuljeongcheakui Jeongae (A Discourse Analysis of Science and Technology During the Park Chung-Hee Era)," 2012, 15.

a competitive edge through industrialization, the government decided the country would have to compete in industries that relied more substantially on bigger, more complex, and more expensive equipment and machinery.

How could a government committed to centralized economic planning promote the private use of higher-end equipment and machinery while also maintaining a demonstrable opposition to communism? One part of the answer was to facilitate the acquisition of such equipment directly from other countries. Most Korean companies during the 1960s in fact obtained their equipment through appropriations from other countries, or what became known as technology transfer. The Park government allocated U.S. aid to this end as much as possible and guaranteed loans to Korean companies made by Japanese and European banks.[26]

The other part was to begin funding domestic research specifically to support the use and improvement of the technologies transferred. Call it research on science and technology, but with this particular normative twist: it focused on existing machinery and equipment and prioritized increasing exports as the main outcome.

Prior to the mid-1960s, the research centers for science and technology that did exist across Korea had been public institutions. According to the law at the time, financial supports for a researcher belonging to public institutions were limited and there was a risk that the activities of the institutions were also managed bureaucratically. The Park government's solution to these problems was to establish independent, non-profit research organizations and then fund them with government research contracts. It actually looked to the Battelle Memorial Institute in the U.S. as a model to appropriate. Founded in 1929 with funding from the estate of industrialist Gordon Battelle (1883–1923), the Battelle Institute had long conducted contract research for the U.S. government. Guaranteeing funding streams gave the government effective control over the contents of its research without directly owning it. The Battelle Institute thus lived simultaneously inside and outside of government. The Park government founded the Korea Institute of Science and Technology (KIST) in 1966 with the same idea in mind.[27]

Funding for KIST actually came largely from the U.S. A year earlier, U.S. President Lyndon Johnson (1908–1973) had offered to help establish independent research centers in Korea. At the time, Korea was maintaining a force of roughly 50,000–60,000 troops in Vietnam to help the U.S. fight its war, ultimately sending 300,000 troops. U.S. support for research centers in Korea would be, in part, compensation for this substantial commitment.

The U.S. was also seeking ways of supporting the Park Chung-hee government and expanding diplomatic ties between Korea and Japan so it could build consistent, lasting diplomatic policies in East Asia. At the time, the Japanese were actively using the Vietnam war to stimulate

[26] Rodrik, "Getting Interventions Right," 1995, 61; You, *Gyeongje Seongjang Sinwha-Ui Heowa Sil (The Truth and Falsity of Economic Growth)*," 2011, 51.

[27] Moon, "Hanguk Gwahakgisul Yeonguso Seollip Gwajeong-Eseo Hanguggwa Migukui Yeok-Hal (Reappraisal of the Establishment of KIST)," 2004, 67

their own economy, building industries to support the war effort.[28] Apart from helping Korea expand its capacity to export, the U.S. had few stakes in the actual research KIST would carry out. Its first director, Dr. Choi Hyung-sub (1920–2004) (Figure 3.4), would later write: "The institute was formed by our needs. Nothing came about by the requests of the U.S."[29]

Figure 3.4: President Park presenting a certificate of appointment to Choi Hyung-sub, President of KIST. Source: Korea Institute of Science and Technology.

The Battelle approach did become a successful model for government-sponsored research organizations in a development-oriented environment. Government guarantees of substantial support enabled the centers to attract Korean researchers who had been scattered abroad. The cost to researchers was to accept, even commit to, President Park's Economy First principle. Dr. Choi related a story about his efforts to attract Korean-born Ph.D. students back from the U.S. in KIST's first year. Some students complained that their research would have to be geared toward expanding industrial production and exports. "You need to do research in fields that will contribute to companies," Choi remembered saying, "even if it is not interesting academically." "Wouldn't it be worse,"

---

[28] Moon, "Hanguk Gwahakgisul Yeonguso Seollip Gwajeong-Eseo Hanguggwa Migukui Yeok-Hal (Reappraisal of the Establishment of KIST)," 2004, 63.

[29] Choi, *Bul-I Kkeojiji Anhneun Yeonguso* (*The Research Institute*), 1995, 70.

he continued, if you misunderstood now and quit later? I am trying to obtain a commitment that you will be okay working in this environment when in Korea."[30]

Efforts to make Korea a source of science and technology research for industry concentrated in the Ministry of Science and Technology. Established in 1967, the Ministry gained financial support from the Science and Technology Promotion Law passed that year. The Ministry's main job was to take charge of defining and administering research programs that would stimulate and support industrial expansion. There was no shortage of researchable questions: What sorts of research would be required to reproduce machinery that had originally been produced elsewhere? How could foreign-made machinery fit Korean personnel and practices? How might Korean researchers adapt machinery made for one purpose to other, new purposes? How might researchers improve on the capabilities of available machinery?

Just as the Economic Planning Board had begun promoting vocational secondary education by creating a classification system for the Korean workforce in 1962, so the Ministry of Science and Technology sought to promote higher technical education by categorizing the science and technology workforce in 1969. Most importantly for our purposes, the Ministry championed a new category of scholar engineer, *gwa-hak-gi-sul-ja*, which we translate as "scientist-engineer." Note the character *gwa-hak* added to *gi-sul-ja*, the Board's original term for engineer. *Gwa-hak* was not new. Just as the character *gi-sul* had traveled from Japan to Joseon during the 1890s, reportedly translated by Nishe Amane (1829–1879) to mean "technology/machinery," so it had been with *gwa-hak*, translated by Amane to mean "science/research."

Now decades later, the Ministry of Science and Technology appropriated the term to explicitly add an identity for engineers fully worthy of the status that *yangban* scholar-officials had long enjoyed. As the Ministry formalized the definition a few years later, the *gwa-hak-gi-sul-ja* were technical literati who performed and led creative activities. They were in charge of research and development, planning, and management after graduating from a four-year science and engineering school."[31]

The requirement of graduation from a four-year science and engineering school labeled them as techno-national scholars in ways that neither the Board's original category of engineers, *gi-sul-ja*, nor the subsequent legal category of working scholar-engineers, *gi-sul-sa*, could claim.

We call them scientist-engineers because their primary responsibilities lay in performing and managing research. Combining the identities of engineer and scientist made sense for the period. Dr. Choi addressed one of the reasons earlier: research was for solving the problems of industry and advancing specific export industries. There was no interest across Korea during the 1960s in what the U.S. National Science Foundation was then championing as basic research. Korean research was

---

[30] Choi, *Bul-I Kkeojiji Anhneun Yeonguso* (*The Research Institute*), 1995, 60.

[31] Ministry of Science and Technology, *Gwahakgisul Yeongam* (*Science and Technology Annals*), 1973, 72.

directional in the sense that it sought to develop and improve practices of industrial production with the ultimate goal of increasing exports.[32]

A second reason was that scientist-engineers remained few in number. Those who received higher education in engineering formed a class together with those who received higher education in, say, chemistry or physics. The key to scholarly status was completing higher education, and not many had the opportunity to do that.

Scientist-engineers did not carry out research alone, of course. The additional categories set in 1969 carried forward the earlier categories of technicians, *gi-sul-gong*, and craftsmen, *gi-neung-gong*. Those men who worked in and around laboratories supported the research of scientist-engineers. All responded, either explicitly or implicitly, to President Park's motivating threat, "Export or Die!" Women workers had to fight to gain admission to these elite spaces. Few proved successful.

In sum, the Park government's efforts to expand the supply of technical personnel for industry during the 1960s focused primarily on technical secondary education for light industry. Labels for engineers and working scholar engineers were available but not much used on the job. The government's interest in shifting from import substitution to export-led growth led it to create the Ministry of Science and Technology and seek ways of expanding domestic research on machinery and equipment. One way was to champion the image of the scientist-engineer and hope students would seek higher technical education in order to claim it.

RESISTANCE IN THE "SECOND ECONOMY"

President Park Chung-hee was all too aware of the challenges he faced in persuading Koreans who valued moral development and education that "Economy First" should somehow become a guiding slogan. The economy had never been first.

During his first two years as a civilian leader, Park appealed to the literate Korean public through two books. The significance of these books cannot be overstated given his own background. Park Chung-hee had been born into a poor peasant family in a remote rural province, which basically limited his pathway to leadership to the military. His heavy reliance on military officers even in his civilian government and explicit rejection of old leaders from the Rhee Syng-man regime made evident his distrust of elite families, the old *yangban*.

Park's first book, *Our Nation's Path: Ideology of Social Reconstruction*, appeared in 1962. It offered a powerful moral argument for overcoming differences in status, and instilling what he called a collective sense of "national self-consciousness." Such required eliminating what he called "privilege consciousness" and "factionalism." These he traced back to the "diseases of the social structure" during the *Yi* Dynasty, the Joseon period. Control over the political economy of *Yi* society, he argued, had lain in the hands of the landlord-bureaucrat class. Its hierarchical structure

---

[32] One of us, Gary Downey, defines the "directionality" of engineering knowledge and work as the materialization of normative commitments in concrete techniques and practices (Downey, "Foreword," 2013).

had inhibited independent spirit, pioneering spirit, and enterprising spirit. But the past was also filled with bitter struggles overcoming adversities from which Koreans could draw inspiration. The invention of Korea's distinctive *Hangeul* script had demonstrated strength and independence. Practices of local autonomy had provided sources of new initiative: indigenous literature explored the lives of ordinary people; pragmatic philosophies challenged rigid hierarchy; and many generations of Koreans resisted outside invasions.[33]

In Park's vision, the biggest challenges to Korea lay not in the communist threat but in redirecting local commitments away from old privileges toward what he called new "social reconstruction." Social reconstruction would be built on "promotion of the public welfare, freedom from exploitation, and the fair distribution of income among the people."[34]

The means for achieving social reconstruction would be technological and economic. Under Park's leadership, Koreans would make sure that "our poor economic power is greatly strengthened" through sound, long-range planning. It would stimulate creativity and the "spontaneity of private enterprise." Park concluded with a direct appeal to educated male elites. He asked "scholars, educators, scientists, artists, and other men of culture" to be the "vanguard of those responsible for the reconstruction of our fatherland."[35] With communism in the north explicitly promising new power to the peasantry, one can see how Park's vision could have appealed to men who had long been disenfranchised. In the wake of land reform, the book sought to convince those men (and parents) who could read that further yielding power and turning their attention to the export economy could help build a new kind of collective, techno-national whole.

The second book was *The Country, The Revolution, and I*, mentioned at the outset. Published a year later, it took a more pragmatic approach to persuasion. Demonstrating Park's passion for numbers, it painstakingly documented the success of the Economy First principle during the regime's first year. The five-year plan envisaged, for example, an ambitious annual growth rate in manufacturing of 15%. By the second year, Park reported, it had already reached 11.5%. And ongoing investments in the integrated iron mill as well as manufacturing facilities for diesel engines, textile machines, electrical appliances and cables, quick-freezing fishery products, automobiles, fertilizer, oil refining, and cement would make it increase even faster.[36] We will not attempt here to do justice to the remarkable level of detail in this book. Its message was intense optimism, buoyed by a sea of numbers.

This second book generated comparative lessons from revolutionary action by Sun Yat-sen in China, mid-level *samurai* in Japan's Meiji reform, Mustafa Kemal in Turkey, and Gamal Abdel

---

[33] Park, *Our Nation's Path*, 1970, 27, 22–27.

[34] Park, *Our Nation's Path*, 1970, 213.

[35] Park, *Our Nation's Path*, 1970, 214, 234; Hahm and Plein, "Institutions and Technological Development in Korea," 1995, 60–61.

[36] Park, *Gugga, Hyeogmyeonggwa Na* (*The Country, the Revolution, and I*), 1970, 80, 81–84.

Nasser in Egypt. It drew particular inspiration from the rapid redevelopment of Germany after WWII. Park found it intuitively obvious that Koreans had to establish a "self-supporting economy and industrial revolution" in order to experience collective "renaissance and prosperity."[37]

Six years into his regime, President Park spoke to graduates of Seoul National University who had worked hard as children to pass the entrance examination, had now completed their studies, and would soon be carrying the status of cultivated, moral, educated scholars into elite positions in government or business. Once again he turned to his distinctive image of industrial production for exports. Past leaders had "turned their backs on the term production," he said. Colonization by the Japanese had left behind a "bureaucratic attitude," "wasteful habits," and a "negative physiology of unconditional defiance." The new, developing Korea needed "leader[s] for production" who would be "leader[s] for development."[38]

Marching in step, the Supreme Council and Economic Planning Board highlighted the value of labor at every turn. They supported policies with a proliferation of moral slogans, such as: "Let's work like our life depends on it"; "Work night and day on a twenty-four-hour working system"; "If we set our hearts to it, we can achieve it"; and "Let's not lose this opportunity and drive it through."

Sloganeering became active socialization when President Park issued the National Charter of Education in 1968 (see Figure 3.5). He modeled this document and practice on the 1890 rescript on education issued by the Japanese emperor.[39] Children of the Japanese empire had recited the rescript aloud every day at school until the end of the Pacific War. Now in Park's seventh year, Korean children would recite aloud his own moral, economic charter. A North Korean commando attack on the President's residence, the Blue House, justified its urgency.

"Patriotism and nationalism committed on an anti-communist national spirit," the Charter asserted, "is our way of life and the foundation to realize the ideas of free society." Building this way of life depended on reconstruction and renewal: "We were born on this land with the historic mission to revive the nation" by "resuscitat[ing] our ancestor's proud spirit." Yet the future must not be simply a continuation of the past. "Let us as diligent citizens armed with confidence and pride," it implored its reciters to accept, "… collect our nation's wisdom and continuously make an effort to create a new history."[40]

---

[37] Park, *Gugga, Hyeogmyeonggwa Na* (*The Country, the Revolution, and I*), 1970, 171.

[38] The Presidential Secretariat, *Park Chung-Hee Dae-Tong-Lyeong Yeon-Seol-Mun-Jib 2: 1963. 12-1967. 6 (Speech by President Park Chung-Hee: 1963. 12-1967. 6)*, 1973, 957.

[39] A rescript is a response to a call.

[40] Ministry of Education, *Gukmingyoyukheonjang Dokbon* (*National Charter of Education Reader*), 1968, 1–2.

Figure 3.5: Declaration of the National Charter of Education, December 5, 1968. Source: National Archives of Korea.

The policy slogans, National Charter, and extensive use of children to promote government programs were all part of what President Park called the "second economy." The second economy was a moral economy. Initiatives within it focused on "spiritual" development, alongside the material developments of industry that became the first economy.[41] In his annual inaugural press conference held in 1968, President Park asserted that the "'second economy' is not academic terminology … or an academic concept, … [but] a term that I called to mind myself." Its goal was to hone in on the issue of acceptance, naming the attitudes and philosophical stance he sought himself to embody, and appropriating the neo-Confucian flow from spiritually disciplining the self to materially fulfilling one's duties:

> The meaning of the term is that "economic construction" or "modernization movement" can be efficiently accomplished only when we have decent mental attitudes as well as

[41] Seth, *Education Fever*, 2002, 204.

physical efforts in our endeavor to construct the economy. What we have meant here is that if we can label the so-called conventional concept of economy, including production increase, export, and construction, as first economy, then we can label such invisible mental aspects, or the philosophical base or platform of modernization as second economy.[42]

In practice, the disciplining of self in Park's second economy meant increased military training in primary and secondary curricula as well as systematic indoctrination in the values of loyalty to the state, anti-communism, and intense nationalism. The Ministry of Education standardized anti-communist texts in the early 1960s. *The Road to Achieving Unity through Victory over Communism* aimed at middle school readers, and *The Road to Safeguarding Freedom* was for high schools.[43]

The material expectations of duty and discipline permeated the large-scale projects. President Park involved himself personally, for example, in all phases of the construction process for the Gyeong-bu Expressway (see Figure 3.6). Beginning in 1968 and designed to span 428 kilometers linking Seoul with Busan, the project was completed in two and a half years. The rush to finish struck many as a military-style charge. Indeed, the army assigned construction supervisors, who, in turn, pressured workers with extra duties and overtime work.[44]

One construction worker later described a moment when his crew faced the challenge of removing a hill to make way for the expressway. "Captain Noh Boo-woong," he said, "cried out the command, 'Our enemy is that *Da-ri-nae* hill!'" Unfortunately the ground was frozen, and bulldozers could make no headway. Furthermore, Captain Noh had no knowledge of how to build a highway. However, crew members found motivation in his passion, continued the worker, and he "spurred our morale to win." The crew won its battle against the hill with a good military strategy: they blew it up with TNT.[45]

To find funds for large projects, President Park worked on both the first and second economies. The International Bank for Reconstruction and Development, founding institution in the World Bank, had denied the highway project a loan on the grounds the country lacked the technical capability for construction. The Park government pressed ahead, raising funds internally by increasing the tax on fuels and by applying war compensation from Japan that Korea had recently received. Upon its completion, the government used the successful construction to remind Koreans of the

---

42 The Presidential Secretariat, *Park Chung-Hee Dae-Tong-Lyeong Yeon-Seol-Mun-Jib 2: 1963. 12-1967. 6 (Speech by President Park Chung-Hee: 1963. 12-1967. 6)*, 1973, 133–134. In an official history published in 1988, the Ministry of Education defined the second economy as "not material modernization" but as providing "a firm spiritual basis for modernization through a revival of the national [*minjok*] spiritual education" (Ministry of Education, *Gyo-Yuk 40 Nyeonsa (Forty-Year History of Education)*, 1988, 249.

43 Han, "*Yushin*-Cheje Bangong Gyogyuk-Ui Silsang-Gwa Yeonghyang (The Reality and Impact of Anti-Communist Education During the *Yushin* Regime)," 1997, 336; Seth, *Education Fever*, 2002, 209.

44 National Institute of Korean History, *Geunhyeondae Gwahaggisulgwa Salmui Byeonhwa (Science, Technology, and Changing Life in Modern Times)*, 2005, 259.

45 Korea Expressway Corporation, *Ttamgwa Nunmului Daeseosasi: Gosokdoro Geonseolui Bihwa (An Epic with Sweat and Tears)*, 1980, 92–93.

suffering the Japanese had inflicted on them and define it as material evidence of their collective commitment to economic development.

Figure 3.6: Burning incense at the opening ceremony of Gyeong-bu Expressway.[46] Source: National Archives of Korea.

Increased production, successful import substitution, and the expansion of exports did in fact take place. The average annual growth rate for the manufacturing industry from 1964 to 1971 was an amazing 21%.[47] Total exports increased from 100 million USD in 1965, or 5.6% of gross domestic product, to 1.1 billion USD in 1971, or roughly 12%.

For our purposes, however, there was a big however. The commitment to, even love for, vocational and technical education that President Park articulated so fervently and persistently did not scale up across the Korean population as he had wanted. It did not become techno-national formation. And technical work did not in fact gain new, high status. Remember that the Japanese had forced Koreans away from academic education and into technical education that led to low-

[46] In this photo, the President and First Lady are expressing their condolence to the seventy-seven people who died during the construction of Gyeongbu Expressway.

[47] Kim, "1960 Nyeondae Suchuljihyangjeok Gongeophwa Chujin (Export-Oriented Industrialization Policy in the 1960s)," 2005, 285.

level jobs. After the Pacific War, the pressure burst for educational pathways that would "lead to … prestigious career[s] in government or business or simply confer the high status of a scholar," akin to the old Joseon scholar-official.[48]

The research on this subject is rich and complex. We will simply note, following the historian Michael Seth, that a "fever" for academic education that had long been available only to *yangban* men spread across the territory after land reform made many new opportunities available to peasant men and women. The Park government found itself working against powerful new images of opportunity and status scaling up across the country that had nothing to do with increasing exports. During the 1960s, educators continued to place high value on moral and ethical training, as well as comprehensive grounding in history, literacy, math, and general science. Popular commitment to "nonmarket" return on investments in education became "and probably had always been… more significant in promoting educational demands."[49]

More than defeating communism or competing with powerful capitalist countries, the dominant pressure many families felt at different income levels pertained to education. Both mothers and fathers wanted to get their sons and, to an increasing extent, daughters into primary and secondary schools that could lead to higher education in universities. In 1962, the Ministry of Education attempted to quell this surge of desire in directions other than technical education by instituting a national qualifying exam for higher education. It started using test scores to limit enrollments. Schools resisted, however, and the government relented. In 1964, individual college entrance examinations replaced the national exam.

Following the commando attack and new National Charter in 1968, the Ministry tried again, attempting to use a national exam prior to the individual exams to elevate vocational and technical education relative to academic education. This time, collaborations of private schools, public schools, and motivated parents successfully thwarted the effort.

By 1966, the number of agricultural, technical, commercial, and fisheries high schools had indeed increased from 283 to 337, but the number of academic high schools had increased from 338 to 397. The ratio between them remained constant. Enrollment at vocational schools increased from 100,000 in 1960 to 474,000 in 1975, but enrollment at academic schools increased from 165,000 to more than 648,000 across the same period.[50]

The bottom line is that advocates of vocational secondary education made little headway in relation to academic education. Five years after it passed, the Industry Education Promotion Act of 1963 was simply shelved. Vocational enrollments at the primary and secondary levels were

---

[48] Seth, *Education Fever*, 2002, 112.

[49] Ministry of Education, *Gyo-Yuk 50 Nyeonsa* (*Fifty-Year History of Education*), 1998; Seth, *Education Fever*, 2002, 252, 248, 114.

[50] Ministry of Education, *Tonggyero Bon Hanguk Kyo-Uk-Ui Baljachwi* (*The Evolution of Korean Education as Seen through Its Statistics*), 1997, 87; Seth, *Education Fever*, 2002, 121, 128.

mostly those who could not pass academic entrance exams, primarily lower-class and rural boys and, sometimes, girls.

The one area of technical education that did expand disproportionately was higher education in engineering and the applied sciences for prospective scientist-engineers. This was not only because the Park government appealed to scientist-engineers to actively cooperate with and lead its mission to industrialize, which it did do.[51] It was also because higher education even in technical areas warranted the image of the scholar and status as technical literati. Even scientist-engineers could be cultivated, morally complete, and worthy of career pathways leading to elevated positions in government or business.

In 1962, 7,685 students entered higher education in science and engineering, actually more than the 7,010 who entered in the humanities and social sciences. A decade later, the first had grown to 19,300 and the second only to 13,020. Yet we must not be misled. Despite these numbers, science or engineering was nearly always not the first choice of a student who passed the relevant admissions examinations. Technical fields, even at the highest levels, were still for lesser students.

REFERENCES

Bowker, Geoffrey C. and Susan Leigh Star. *Sorting Things Out: Classification and Its Consequences.* Cambridge, Mass.: MIT Press, 1999. DOI: 10.3395/reciis.v4i5.424en. 57

Choi, Hyeongseop. *Bul-I Kkeojiji Anhneun Yeonguso (The Research Institute: Its Light Never Goes out).* Chosun Ilbo Co., 1995. 64, 65

Downey, Gary Lee. "Foreword." In *Engineering Practice in a Global Context: Understanding the Technical and the Social*, edited by Williams, Bill, et al., London, UK: CRC/ Balkema Taylor and Francis., 2013. DOI: 10.1080/19378629.2010. 519772. 66

Economic Planning Board. *Gwa-Hag-Gi-Sul Baeg-Seo (Science and Technology White Paper).* Seoul: Economic Planning Board, 1962. 58

Graham, Loren R. *The Ghost of the Executed Engineer: Technology and the Fall of the Soviet Union.* Cambridge, Mass.: Harvard University Press, 1993. 59

Hahm, Sung Deuk and L. Christopher Plein. "Institutions and Technological Development in Korea: The Role of the Presidency." *Comparative Politics* 28, no. 1 (1995): 55-76. DOI: 10.2307/421997. 67

Hahn, Young-Whan. "Administrative Capability for Economic Development: The Korean Experience." <manuscript>, 1995. 55

---

[51] Korea Federation of Science and Technology Societies (KOFST), "Gwahak *Yushinui* Bangan (The Way of Science Restoration)," 1973; Moon, "Park Chung-Hee Sidae Damhwamuneul Tonghae Bon Gwahakgisuljeongcheakui Jeongae (A Discourse Analysis of Science and Technology During the Park Chung-Hee Era)," 2012.

Han, Man-gil. "*Yushin*-Cheje Bangong Gyogyuk-Ui Silsang-Gwa Yeonghyang (The Reality and Impact of Anti-Communist Education During the *Yushin* Regime)." *Yoksa Bipyeong* 38, no. (1997): 333-347. 70

Han, Seung-jo. "Hanguk Jeongchi Elit-Ui Chungwon Yuhyeong (The Recruiting Pattern for Korean Political Elites)." *The Journal of Asiatic Studies* 18, no. 2 (1975): 1-47. 56

Hyun-mee, Kim. "Hangugui Geundaeseonggwa Yeoseongui Nodongkwon (Modernity and Women's Labor Rights in South Korea)." *Journal of Korean Women's Studies* 16, no. 1 (2000): 37-64. 63

Jeon, Sang-geun. *Hangug-Ui Gwa-Hag-Gi-Gul Gaebal: Han Jeong-Chaek Ibanja-Ui Jeung-Eon (Science and Technology Development in Korea: A Policy Maker's Testimony).* Seoul: Sam Gwa Kkum, 2010. 61

Kim, Hyung- A. *Korea's Development under Park Chung Hee: Rapid Industrialization, 1961–79.* London ; New York: RoutledgeCurzon, 2004. 57

Kim, Kwang-seok. "1960 Nyeondae Suchuljihyangjeok Gongeophwa Chujin (Export-Oriented Industrialization Policy in the 1960s)." In *Hanguk Geundaehwa, Gijeogui Gwajung (Modernization of the Republic of Korea, A Miraculous Achievement)*, edited by Cho, I-je and Carter Eckert, Seoul: Montly Chosun, 2005. 71

Korea Expressway Corporation. *Ttamgwa Nunmului Daeseosasi: Gosokdoro Geonseolui Bihwa (An Epic with Sweat and Tears: A Secret Story of the Highway Construction).* Seoul: Korea Expressway Corporation, 1980. 60

Korea Federation of Science and Technology Societies (KOFST). "Gwahak *Yushinui* Bangan (The Way of Science Restoration)." *Science and Technology* 44, no. 1 (1973): 7-8. 73

Lee, Ki-hong. *Gyeong-Je Geundaehwa-Ui Sumeun Yiyagi (Hidden History of Economic Development).* Seoul: Bois-sa, 1999. 57

Lee, Sang-chul. "Park Chung-Hee Sidae-Ui San-Eob-Jeong-Chaek (Industrial Policy of the Park Chung-Hee Era)." In *Gae-Bal-Dog-Jae-Wa Park Chung-Hee Sidae*, edited by Lee, Byeong-cheon, Seoul: Changbi Publishers, 2012. 61

Lee, Won-ho. *Sireop Gyo-Yuk (Industrial Education).* Seoul: Hawoo, 1996. 61

Ministry of Education. *Gukmingyoyukheonjang Dokbon (National Charter of Education Reader).* Seoul: Ministry of Education, 1968. 69

Ministry of Education. *Gyo-Yuk 40 Nyeonsa (Forty-Year History of Education).* Seoul: Ministry of Education, 1988. 70

Ministry of Education. *Tonggyero Bon Hanguk Kyo-Uk-Ui Baljachwi (The Evolution of Korean Education as Seen through Its Statistics)*. Seoul: MOE, 1997. 73

Ministry of Education. *Gyo-Yuk 50 Nyeonsa: 1948–1998 (Fifty-Year History of Education: 1948–1998)*. Seoul: MOE, 1998. 72

Ministry of Science and Technology. *Gwahakgisul Yeongam (Science and Technology Annals)*. Seoul: MOST, 1973. 65

Moon, Man-yong. "Hanguk Gwahakgisul Yeonguso Seollip Gwajeong-Eseo Hanguggwa Migukui Yeok-Hal (Reappraisal of the Establishment of Kist: Roles of the U.S. And Korea)." *Journal of the Korean History of Science Society* 26, no. 1 (2004): 57-86. 64

Moon, Man-yong. "Park Chung-Hee Sidae Damhwamuneul Tonghae Bon Gwahakgisuljeongcheakui Jeongae (A Discourse Analysis of Science and Technology During the Park Chung-Hee Era)." *Journal of the Korean History of Science Society* 34, no. 1 (2012): 75-108. 60, 63, 73

National Institute of Korean History. *Geunhyeondae Gwahaggisulgwa Salmui Byeonhwa (Science, Technology, and Changing Life in Modern Times)*. Seoul: Doosan Dong-A, 2005. 70

O, Won-chul. *Park Chung-Hee-Neun Eotteogge Gyeongjaeganggug Mandeul-Eotna (President Park Chung-Hee's Leadership and the Korean Industrial Revolution)*. Dongsuhbook, 2010. 59, 62

Park, Chung-hee. *Gugga, Hyeogmyeonggwa Na (The Country, the Revolution, and I)*. Seoul: Hollym Corp., 1970. 53, 68

Park, Chung-hee. *Our Nation's Path: Ideology of Social Reconstruction*. Seoul: Hollym Corp., 1970. 54, 67

Park, Gil-sung and Kyung-pil Kim. "Park Chung-Hee Sidae-Ui Gugga-Gi-Eop Kwangae-E Daehan Jaegeomto: Gi-Eob-Eul Bunseog-Ui Jungsim-Euro (Rethinking State-Business Relations in Park's Era: Business on the Center Stage)." *The Journal of Asiatic Studies* 53, no. 1 (2010): 126-154. 55, 62

Park, Jin-geun. *Hanguk Yeogdae Jeong-Gwonui Juyo Gyeongje Jeongchaek (Major Economic Policies of the Korean Government)*. Seoul: KERI (Korea Economic Research Institute), 2009. 59

Porter, Theodore M. *Trust in Numbers: The Pursuit of Objectivity in Science and Public Life*. Princeton, N.J.: Princeton University Press, 1995. 57, 59

Rodrik, Dani. "Getting Interventions Right: How South Korea and Taiwan Grew Rich." *Economic Policy* 20, no. (1995): 55-83. 55, 63

Seth, Michael J. *Education Fever: Society, Politics, and the Pursuit of Schooling in South Korea*. Honolulu, HI: University of Hawai'i Press, 2002. 55, 61, 69, 70, 72, 73

The Presidential Secretariat. *Park Chung-Hee Dae-Tong-Lyeong Yeon-Seol-Mun-Jib 2: 1963. 12-1967. 6 (Speech by President Park Chung-Hee: 1963. 12-1967. 6)* Seoul: The Presidential Secretariat, 1973. 68, 70

You, Jong-il, ed. *Gyeongje Seongjang Sinwha-Ui Heowa Sil (The Truth and Falsity of Economic Growth).* Seoul: Sisa IN Book, 2011. 55

CHAPTER 4

Engineers for Heavy and Chemical Industries: 1970–1979

In the early 1970s, President Park Chung-hee presented banners to select secondary schools, bearing an honorary title he had created himself: "The Engineer is Bearer of the Nation's Industrialization." Engineering high schools for boys were extremely challenging. In order to arrive on site in time, the students had to rise at the break of dawn. Before heading off to training halls for classwork, they participated in the daily chant "Just do it! We can do it!" They would not then return home until after nightfall. During the final six months of a three-year enrollment, they were to devote more than seven hours per day in practical training for specific industries. The students were preparing to become industrial technicians, a category of practitioner that the Park administration was filling with new meaning.[1]

In 1973, the Park government announced a new emphasis on developing heavy and chemical industries, especially those producing steel and heavy equipment, as well as large-scale chemical plants. It was shifting the headlong drive for exports away from light industries into industries that required higher levels of technical expertise and larger-scale equipment, machinery, and manufacturing processes. According to the government, these would not only achieve economic prosperity for Koreans but also elevate the country to international power and prominence. They would create a new Korea.

The engineering schools would support this effort. President Park encouraged the students who attended to embrace his new program and come to see themselves as techno-national leaders, working on behalf of the country as a whole. At Busan Mechanical High School for boys, the banner was mounted on a tower in 1975 (Figure 4.1). The phallic shape of the tower topped by an arrowhead suggested a missile ready for take-off. It graphically linked the advancement of Korea to rapid industrialization by means of special boys.

---

[1] O, *Park Chung-Hee-Neun Eotteogge Gyeongjegangguk Mandeureonna* (*President Park Chung-Hee's Leadership and the Korean Industrial Revolution*), 2010, 424.

Figure 4.1: President Park's Banner at Busan Mechanical Engineering High School. Source: http://gomoonok.com.

Two years earlier, President Park had effectively redefined himself as a development dictator. He declared a national state of emergency, arrested critics and opponents of his policies, and issued a new constitution that gave him power for life. He called the power grab *Yushin* (유신), meaning restoration, following the Japanese term for the Meiji Restoration of 1871 that had granted the emperor political power for life. No longer directly accountable to an electorate, Park could promulgate laws without ratification by the National Assembly. His government could push hard to develop heavy and chemical industries, including educational initiatives to support them.

Park's self-coup also grounded a new emphasis on military operations. As is the case with every episode in this book, the situation was more complex than we can present it. But it is important to point out here that the U.S. government played a role. In 1969, U.S. President Richard Nixon announced that the U.S. would no longer fight wars on the soil of allied countries. It would provide instead a nuclear umbrella and sell allies conventional weapons to build up their own military capabilities. Korea's new, 1973 emphasis on heavy and chemical industries included an institutional innovation that was already well-known in the U.S.—a defense industry. As in the U.S., the Korean defense industry would produce weapons and other military resources, operating nominally in the private sector but funded by government contracts.

As we will see, however, by the end of Park Chung-hee's rule in 1979, the cracks in the *Yushin* dictatorship had become obvious, even for our objects of study. Not even the absolutist, often brutal, exercise of centralized authority backed by claims of national emergency would substantially upgrade the status of vocational secondary education, technical higher education, and technical practitioners all at levels, including the scholarly scientist-engineers. Park would not get the person he tried to create, the technical practitioner who embodied industrial Korea as a whole.

MAKING HEAVY INDUSTRY KOREAN

In 1970, the Park government assigned the Hyundai Group the task of constructing a shipbuilding yard. Note how the verb "assigned" indicates a blurred boundary between the evolving public and private sectors. The government would not build the shipyard itself. But since government officials selected which industry group would do the work without competitive bidding, the shipyard clearly became a government project as well as a private one. Consider also financial support. When the Japanese government denied a loan to the Hyundai Group to fund construction, the Park government stepped in to ensure funding for the project.

Economists interested in the political economy of this period tend to highlight the active participation of the Korean government in developing the heavy and chemical industries.[2] For our purposes, these collaborations are interesting for another reason: they constituted efforts by the Park government to add a Korean identity to the heavy and chemical industries, effectively making them Korean. To the extent the projects were successful, so could the contributing participants, including technical workers at all levels, be able to claim to be working for Korea as a whole.

In order to acquire the necessary expertise to design the shipyard and supervise its operations, Chung Ju-yung (1915–2001) of the Hyundai Group traveled to Scotland in 1971 to seek support from Appledore Shipbuilders (Figure 4.2).

[2] Hahn, "Administrative Capability for Economic Development: The Korean Experience," 1995; Kuznets, "Government and Economic Strategy in Contemporary South Korea," 1985; Hughes, "Why Have East Asian Countries Led Economic Development?," 1995.

Figure 4.2: Chung Ju-yung. Source: Hyundai Motor Company.

The Appledore chairman was reluctant to make a long-term commitment to a Korean company that had no experience in shipbuilding, nor any orders in hand. As the story goes, Chung then pulled out a 500 KRW note (Korean won, pegged to the USD, about 1.5 dollars in 1971). The note depicted the Joseon Dynasty hero Admiral Lee Soon-shin (1545–1598) and the battleship *Geo-buk-seon* (거북선 turtle ship, because of its shape) he had ordered to be built (Figure 4.3). It was the world's first iron clad ship, used to fight the Japanese. Chung reportedly told the Appledore chairman, "Korea was making ironclad ships in the 1500s, 300 years before the British."[3] He got the contract. Chung also went on to secure orders for two ships from a Greek shipping tycoon. The Hyundai Heavy Industry Company was formally established in 1973. It launched the two tankers a year later (Figure 4.4).

With further support from the government, the Samsung and Daewoo groups followed suit and started building shipyards. Over the next decade, the shipbuilding industry became wholly established by means of creative public-private collaborations. Forty years later, Korea had six of the top ten shipyards in the world, and held first place in the number of outstanding orders. Shipbuilding had become Korean.

---

[3]   Hong, *Lee Byung-Chul Dae Chung Ju-Yung* (*Lee Byung-Chul Vs. Chung Ju-Yung*), 2011, 169.

Figure 4.3: The 500 KRW note circulating in 1971. Source: Bank of Korea, Money Museum.

Figure 4.4: 1974 Launching ceremony of Atlantic Baron (260,000 ton oil tanker) by Hyundai Shipbuilding Company. Source: National Archives of Korea.

Since the early 1960s, the Park government had provided economic support and incentives to those family-owned businesses that promised rapid growth in exports. When he first assumed control of the government, Park had been strongly suspicious of wealthy industrialists, some of whom were former *yangban* elites. In his judgment, many had been too close to the Rhee Syngman government, corrupting its administration. Indeed, Park dismissed 35,000 of the country's 240,000 civil servants when he first took power.[4] Over time, however, the government's dual commitment to fighting communism and expanding exports led the Park government back to the family-owned businesses.

Figure 4.5: President Rhee Syng-man. Source: The memorial association for President Rhee Syngman.

During the 1960s, the larger companies had focused on producing textiles and wigs. By 1970, many were becoming conglomerates that a decade later would be labeled "*chaebol* (재벌)." A *chaebol* was a multi-industry group of connected companies, linked by common family ownership. Unlike their Japanese counterparts, the *keiretsu*, Korean conglomerates did not include banks and so were dependent upon external financial support for major initiatives. The Park government ensured such financial support by mediating arrangements for financing from foreign banks, guaranteeing loans from both foreign and domestic banks, and providing direct financial support itself.

---

[4] Lie, *Han Unbound*, 1998, 53.

By the 1970s, the *chaebols* were large enough to enact the new, ambitious directives emanating from the Blue House[5] and Park administration to develop heavy and chemical industries. Between 1974 and 1979, for example, the government's National Investment Fund provided nearly 1.5 trillion KRW in loans to industry. Of this, nearly 900 billion KRW, or more than 60%, went to heavy and chemical industries.[6]

Lacking the material resources to develop advanced heavy and chemical industries from scratch, the aggressive Park government focused on helping companies license technologies from firms in other countries. With increasing exports from light industries fueling cash reserves, the government was able to help individual companies import entire factories on a turnkey basis.[7] The image of the turnkey factory was attractive—you just needed to turn the key to start it up. Men workers could quickly learn how to operate new factories from guest workers brought in to complete the addition of industrial identities to Korean workers. Importing factories in part or in whole reduced the amount of time necessary to achieve normal operation and, hence, the financial risk of such large investments.

The most prominent example of the government making heavy industry Korean was the huge iron and steel complex in the southeast. Because the Pohang Iron and Steel Company, or POSCO, would provide iron and steel to industries all across the country and the project was both hugely expensive and risky, the state took 75% ownership. It allocated 25% ownership to the Korea Tungsten Company. Funding came primarily from Japan. POSCO developed its workforce both by sending Korean workers overseas for training, especially to Japan, and by importing skilled, frequently retired, male Japanese technicians as guest instructors. One Japanese technician later reported his pleasure at the attention he received: "Because we always had POSCO staff members with note pads around us, each of us felt like a cult leader."[8] Steel production in Korea increased from 500,000 tons in 1970 to 8.5 million tons a decade later (Figure 4.6). Thirty years after that POSCO was the world's #1 steel corporation.[9]

---

[5] The Blue House was the presidential residence, akin to the White House in the U.S.

[6] Lee, "Park Chung-Hee Jeongbu-Ui Gyeongje Jeongchaek (Economic Policies in the Park Chung-Hee Era)," 2009, 97.

[7] Kim, "Gwahakgisul Jinheung-Gwa Gyeongje Baljeon (The Promotion of Science, Technology, and Economic Development)," 2005, 357.

[8] History, *Geunhyeondae Gwahaggisulgwa Salmui Byeonhwa (Science, Technology, and Changing Life in Modern Times)*, 2005, 198.

[9] Pohang Iron and Steel Company (POSCO), *Posko 35 Nyeonsa (35-Year History of Posco)*, 2004, 48.

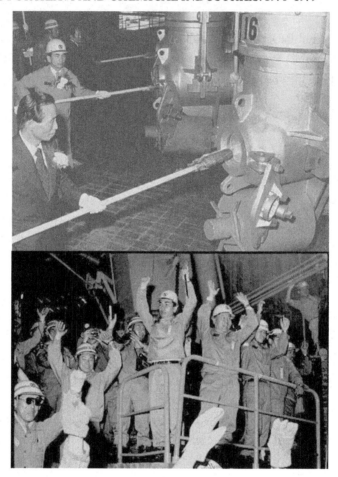

Figure 4.6: The 1976 ignition ceremony of POSCO's second blast furnace. Source: National Archives of Korea.

The government also actively encouraged companies in potential export industries to reverse-engineer popular products manufactured in other countries. The first Korean home appliance company, LG, for example, hired German technicians skilled in electronics. Working in a tiny storage-sized factory, Korean workers integrating German expertise successfully copied and reproduced major components from Japanese AM radios (Figure 4.7).[10]

---

[10] Kim, *Imitation to Innovation*, 1997, 134.

Figure 4.7: Korean workers in the LG radio factory in 1962. Source: National Archives of Korea.

In addition to facilitating imports of industrial technologies and technical expertise, the Park government sought to ensure that industrial developments achieved within the country stayed in the country. The Technology Development Promotion Act of 1972 protected the domestic manufacturer of a new technology, so it could secure returns on its investment, by restricting the importation or manufacture of similar products. A few years later, for example, the Korean company Sunkyoung (later SK Group) successfully produced the polyester film necessary to make videotapes. Its competitor, Samsung, found a way to gain access to the technology from a Japanese company without any royalties. Dr. Choi Hyung-sub, Minister of Science and Technology (see Figure 3.5 above), denied its request to do so, requiring it to license the technology from its domestic competitor. "Are you telling me," he reported later, "that you wish to throw away the technology we worked so hard to develop? No, that will not happen."[11] Escaping from subordination to Japan required making Korean companies autonomous wherever possible.

Likewise, the Technical Services Support Act of 1973 required all contractors for technical services to be Korean unless they were incapable of providing the services required. The Support of Specific Research Institutes Act, also passed that year, provided endowment support for research institutes specializing in problems in industry. Organized by industry sector, the institutes retained operational autonomy but responded wholly to research requests from the government. Subsequent

[11] Choi, *Bul-I Kkeojiji Anhneun Yeonguso* (*The Research Institute*), 1995, 227.

legislation authorized support for corporate and university-affiliated research centers. Still other laws provided direct research support for specific industries, including the Machinery Industry Promotion Act (1967), Shipbuilding Industry Promotion Act (1967), Electronics Industry Promotion Act (1969), Steel Industry Promotion Act (1970), and Automobile Industry Long-term Promotion Plan (1973). Overall, in the Park government vision, although Korean companies in the heavy and chemical industries operated formally in the private sector, they could also embrace fully Korean identities by fulfilling responsibilities assigned by the Korean government.

During the 1970s, the country-wide rush to develop heavy and chemical industries became a quasi-military action in scope and level of coordination. With the Ministry of Science and Technology reporting directly to the President and the Korea Institute for Science and Technology responsible for innovations defined by the administration, a triangular configuration of military-style leadership emerged. The Blue House was the "headquarters," the Ministry of Science and Technology became the "joint chiefs of staff," and KIST the "think tank, supporting the commander-in-chief."[12] Beyond the government, the autonomous research institutes provided ideas and technology, and the companies manufactured, distributed, and exported. Put in other terms, "a role division system" developed in which "the state was responsible for distributing investment and resources based on economic planning and bank management while the private sector was assigned with manufacturing and marketing."[13]

Finally, for industry leaders, accepting the Park vision of a unified Korea frequently meant appropriating and shifting to industry a once-rural image of leadership drawing on neo-Confucian philosophy. This image blended the authority of male leaders to direct with a responsibility to participate and embody the whole. Industry leaders supported the government's admonition to just do it, discussed earlier, with images of them doing it themselves.

Park Tae-jun (1927–2011), founding leader of POSCO, reportedly involved himself deeply in all aspects of the business, from purchasing equipment to construction management (Figure 4.8). After spending time in the department of mechanical engineering at Waseda University, an elite private institution, he had completed his higher education at the Korean Military Academy. In 1971, construction of a hot rolling mill at the steel plant had fallen three months behind the construction schedule set up for it. Park Tae-jun gave the order to pour a thousand cubic meters of concrete daily, far more than had been achieved up to that point.[14] He personally supervised the cement trucks and stayed on site in his poncho in all kinds of weather. To reach the goal, many laborers had to sleep on-site, and exhausted truck drivers could only take naps on the shoulders of the construction roads.

---

[12]   Kim, "Gwahakgisul Ipgukui Haebudo (Anatomical Chart of a Scientific and Technological State)," 2008, 254.

[13]   You, *Gyeongje Seongjang Sinwha-Ui Heowa Sil* (*The Truth and Falsity of Economic Growth*), 2011, 52.

[14]   Seo, *Chulgangwang Park Tae-Jun Yiyagi* (*Steel King*), 2011 [1997], 289.

Figure 4.8: Park Tae-jun. Source: POSCO.

Lee Young-woo, who worked in the field at the time, later described with wonder what happened the day after they first achieved the thousand cubic meters. Exhausted, they were able only to deposit 450 cubic meters. Park Tae-jun's command came through the telex to "report immediately on the reasons why you were ... short and make up for the shortfall by today with no excuse." "We were naïve," this laborer said. "We forgot how tenacious our CEO Park was ...."[15]

To embrace a formerly rural image of male leadership associated with *yangban* status did not mean that *chaebol* leaders faithfully or uniformly integrated this image into their practices. The stories here are varied and complex. What we can see from those occasions in which industry leaders interpreted themselves more broadly as Korean leaders are tenacious efforts to make the heavy and chemical industries and their practitioners icons of Korea and Korean development.

In sum, while the U.S. government was racing against communism by expanding its defense industry and pursuing space travel to the moon, the Park government also raced against communism. It did so by achieving growth rates in exports of up to 40% per year.

---

[15] Lee, *Sindeullin Saramdeul-Ui Habchang* (*Chorus of the Bewitched*), 1998, 79.

VOCATIONAL GRADUATES FOR SPECIFIC INDUSTRIES

Aggressive changes in educational policies sought to fit the making of people to the new emphasis on heavy and chemical industries. The Industrial Education Promotion Law of 1973 authorized training oriented toward the needs of the new industries. Such targeted techno-national formation would not be easy to accomplish, however.

Over the years, the overriding emphasis on preparing for national academic exams and identities as scholar-officials had led even the vocational schools to begin offering academic subjects in preparation for the exams. Through subsequent policy and legislation, vocational schools were required to devote at least 70% of their class time to vocational subjects. Male students found themselves required to spend two to six months of their first two years training at an industrial site and at least four months of their third year working on the shop floor.

To build industrial identities on the job, the Park government extended public-private collaborations to schools and industrial firms. An Employment Training Centers Law created a "system of licenses for industrial instructors," and a Vocational Training Regulation Law expanded training programs to include industrial sites. The Ministries of Defense, Construction, Communication, and Education, as well as the national railroads, "all operated on-the-job training centers." And under the new legislation, even large private operations were expected to provide training programs.[16]

To help attract boys to vocational secondary schools, the National Technical Qualifications Act of 1973 created new categories of technical worker. It also upgraded existing categories to –sa (사), meaning scholar, by developing formal exams to certify and affix their higher statuses, akin to the civil service exams that had long certified literati as scholar-officials. Craftsmen (yes, men) rose from *gi-neung-gong* (기능공) to *gi-neung-sa* (기능사), yet were required only to complete vocational secondary school. Those who spent a year working as a craftsman or had completed two years of junior college could attain the new category "industrial engineer," *san-eob-gi-sa* (산업기사), after completing its exam. Those who completed two years of work after junior college or four years of a university course could complete the exam warranting the new category "engineer," *gi-sa* (기사).

Tellingly, the next category "master craftsman," *gi-neung-jang* (기능장), was actually higher in status than those of industrial engineer or engineer. Its exam was available to craftsmen who had worked at least seven years and to those graduates who had attended a new type of technical higher education institution, the vocational junior college.

Finally, the top status went to the scholarly professional engineer, *gi-sul-sa* (기술사), upgraded from the earlier status of engineer, *gi-sul-ja* (기술자). One could reach that exam several

---

[16] Seth, *Education Fever*, 2002, 126; Cho et al., *Hangug-Ui Gwahakgisul Il-Lyeok Jeongchaek* (*Review of Science & Technology Human Resource Policies in Korea*), 2002, 172–173.

ways, including seven years of work after university education, nine years after junior college, or nine years of experience in the field.[17]

During the 1970s, the Park administration founded 19 mechanical high schools to produce skilled male technical workers for the machine and defense industries. It established 11 experimental engineering high schools and 12 specialized engineering high schools to create technicians who could be dispatched to Korean companies abroad.[18] At the urging of the Economic Planning Board, a total of 82 vocational high schools received financial support to prepare male students to specialize in machinery, electronics, mechanical engineering, and other areas that matched the skills demanded by targeted industries.

Finally, to make sure vocational secondary schools in fact taught vocational subjects, the Ministry of Education instituted a national technical qualification exam in 1978. All those boys expected to graduate from vocational high schools had to demonstrate their competency in the technical areas they studied before qualifying for a degree.[19]

O Won-chul, who became senior adviser to President Park and leader of both the Heavy and Chemical Engineering Planning Group and the Defense Industry Group from their founding until 1979, later characterized the 1960s as the era of (simple) female craftspersons and the 1970s as the era of (skilled) male craftspersons.[20] But this shift was in vision more than in practice, and the image took time to scale up. By the mid-1970s, women-centered manufacturing still "accounted for 70% of the total export" from the country.[21]

During the 1970s, the entrance rate of girls and young women to high schools increased from 54% in 1970 to above 90% in 1979.[22] And the entrance rate of women to colleges exceeded 22% in 1980. But these girls and women were, by and large, not seeking vocational or higher technical education. In 1980, only slightly more than 1% of students in engineering colleges were women. Some women who had completed only elementary or middle schools continued to flow into low-level factory jobs. But while the middle and high school curricula for boys required study of technology and industry, girls were channeled into homemaking.[23] Those who might have been

[17] National Technical Qualification Act #10336; Kim, *Gukgagisuljagyeok Jedo-Ui Hyeonhwang-Gwa Gaeseon Bangan* (*Current Status and Recent Issues in the National Technical Qualification System*), 2004, 5–6.

[18] Cho et al., *Hangug-Ui Gwahakgisul Il-Lyeok Jeongchaek* (*Review of Science & Technology Human Resource Policies in Korea*), 2002, 154.

[19] Kim, *Hanguk-Ui Gongeopjeongchaek Yeongu* (*Research on the Industrial Education Policy of Korea*), 2001, 118; Seth, Education Fever, 2002, 126.

[20] O, *Park Chung-Hee-Neun Eotteogge Gyeongjeganggug Mandeureonna* (*President Park Chung-Hee's Leadership and the Korean Industrial Revolution*), 2010, 48–49.

[21] Kim, "Hangugui Geundaeseong-Hwa Yeoseong-Ui Nodong Gwon (Modernity and Women's Labor Rights in South Korea)," 2000, 41.

[22] Ministry of Education (MOE), *Gyoyuk Tonggye Yeonbo* (*Statistical Yearbook of Education*), 2000.

[23] Lee, "Gwahak Gisulgwa Yeoseong-Ui Jeongchaek Jaeng Jeom (Policies at Issue for Science-Technology and Women)," 2001, 8.

interested in the construction, automobile, steel, shipbuilding, and fertilizer industries had to fight their way into positions. As O Won-chul put it, the 1970s industrial expansion was for men.

RATIONAL SCIENTIST-ENGINEERS FOR LEADERSHIP

The government's new policy preferences for education also took higher education for men by storm. In addition to the polytechnic institutes that enabled graduates of secondary vocational schools to leapfrog to the status of master craftsman, the Park government also encouraged universities to develop specializations specifically geared for production work in heavy and chemical industries. And for those students seeking advanced opportunities and status in industry and government, it founded the Korean Advanced Institute of Science (KAIS, later changed to KAIST to include technology). KAIS offered graduate-level education in science-technology for research, as well as the opportunity to fulfill their military service requirement through work in industrial companies.

During this period, enrollments in engineering and the applied sciences at universities certainly increased significantly, even disproportionately. Between 1972 and 1978, the total number of mostly men students entering engineering each year increased from 19,300 to 33,035. This increase exceeded the increases of men and women in the humanities and social sciences from 13,020 to 20,915 and in education from 5,010 to 11,835.[24] The rate of increase for engineering majors itself increased during the late 1970s, when the private conglomerates became large operations. The positions offered by *chaebols* provided substantial financial rewards and upward mobility into management roles, making the work less narrowly technical and, hence, more scholarly and warranting higher status.

President Park was perhaps himself most interested in those scientist-engineers, *gwa-hak-gi-sul-ja* (과학기술자), who accepted responsibility to develop new, Korean forms of science-technology, *gwa-hak-gi-sul* (과학기술). To describe the purpose and benefits of research, Park actually tended to use the term "science-technology." Indeed, by the 1970s, even the separate term for science, *gwa-hak* (과학), had become roughly synonymous with science-technology. The link with industry and exports was always implied.[25] The government clearly did not intend to pursue scientific knowledge or discovery for its own sake or to support new philosophies of nature via scientific investigation. Science was scientific research directed to benefit industry, increase exports, and create a new Korea free from external interference. University-level basic science was, in these terms, economically useless.[26]

---

[24] Ministry of Education, *Tonggyero Bon Hanguk Kyo-Uk-Ui Baljachwi* (*The Evolution of Korean Education as Seen through Its Statistics*), 1997, 179; Seth, *Education Fever*, 2002, 129.

[25] Moon, "Park Chung-Hee Sidae Damhwamuneul Tonghae Bon Gwahakgisuljeongcheakui Jeongae (A Discourse Analysis of Science and Technology During the Park Chung-Hee Era)," 2012, 88.

[26] Kim, "Gwahakgisul Ipgukui Haebudo (Anatomical Chart of a Scientific and Technological State)," 2008, 248.

The Park government populated its administrative bureaucracies with male experts in economics, science and technology, education, and foreign affairs. Administering laws and policies to promote industrial development required unprecedented amounts of calculation. During the 1950s, formal administration under President Rhee Syng-man had focused more on managing aid funds for reconstruction than generating new sources of income, and the work of allocation involved substantial political negotiation and strategy. Administrative positions tended to be political appointments, the product of patronage. That made them ripe for corruption.

As we explained in the last chapter, President Park's plan to reduce class privilege had initially relied disproportionately on middle-class military men for administrative positions. But the corresponding emphasis on calculation privileged expert knowledge and professional administrative skills. Public officials were held responsible for levels of private-industry performance in their areas of jurisdiction. They in turn held both public and private constituencies accountable for meeting specified quantitative objectives. Park even established an advisory body of technical experts, again all male, to feed him with ideas and help him prioritize the proposals he received from subordinates. The boundary between public and private initiatives blurred routinely, as Park assigned both the responsibility to build a new Korea.

Technical leaders in the Park government typically had gained experience and expertise in countries with advanced industries, most especially Japan and the U.S. Between 1953 and 1967, nearly 8,000 Korean students studied abroad, nearly all male. More than 6,000 of them, or 86%, went to the U.S. Those who returned tended to find their way into the dominant projects of industrialization. Having added foreign identities, they embodied variations of the economic future the Park government imagined for Korea. By contrast, the flow of foreign-educated Koreans into departments "related to public order and government maintenance such as the department of interior or justice" was "almost nonexistent."[27] Producing police and army officials to maintain political order and repress dissent was handled internally.

O Won-chul later referred to the techno-national experts serving the Park government in leadership positions as "technocrats." In EuroAmerican terms, technocrats are technically trained personnel who gain access to government positions and exercise governmental authority. They construe governance as an array of technical activities. In O Won-chul's sense, Korean technocrats were a new kind of scholar-official who focused on exports. "The technocrats," O explained in a memoir, "always considered the possibility of exports when drawing up plans for a factory or constructing one." And "to be competitive in exports," he said, "they endeavored to grow their factory capacity to a global scale, and even further to become the leading company in the world."[28]

[27] Jung, "5.16 Kudeta Ihu Jisikin-Ui Bunhwa-Wa Jaepyeon (Differentiation and Reorganization of Intellectuals since the 5.16 Coup)," 2004, 165.

[28] O, *Park Chung-Hee-Neun Eotteogge Gyeongjegganggug Mandeureonna* (*President Park Chung-Hee's Leadership and the Korean Industrial Revolution*), 2010, 43.

Technocrats charged with imagining and creating new types of exports had special access to Park. Dr. Choi Hyung-sub, first head of the Korean Institute of Science and Technology, proudly reported that "President Park …came to the research institute once or twice a month to converse with the researchers." His visits "raised our pride and social standing" across the administration. "This influence was unimaginably significant" across a territory in which personal recognition from one's leader had since the Joseon period not only conveyed endorsement but also granted legitimacy (Figure 4.9). KIST researchers mobilized this legitimacy into "a shield when they disagreed with the ministers [charged with implementing new research ideas]."[29]

Figure 4.9: President Park encourages researchers while visiting KIST in 1967. Source: National Archives of Korea.

Likewise, once technical officials secured the President's approval for a particular plan, they could proceed largely as they saw fit. In his memoir, O Won-chul remembered a remarkable project briefing session with the President in 1969. Briefing sessions typically ended with budget discussions. In this case, Ministry of Commerce officials had proposed developing the electrical equipment industry. The President evidently found the step obvious. Rather than questioning finances,

[29] Choi, *Bul-I Kkeojiji Anhneun Yeonguso* (*The Research Institute*), 1995, 67.

he simply asked an advisor if he could implement the plan. "Yes, I will do as was proposed by the Ministry of Commerce" came the reply. The President then simply approved the project. The entire budget was secured and implementation began.[30]

By highlighting numerical accountability, quantification increased the government's power to regulate the rate of implementation and delivery of results. In building the heavy and chemical industries, the Park government made it clear it valued not only labor in general, but speedy labor in particular. Working fast became an important virtue for those producing a new Korea, in part because it separated the country from a slow-moving past.

The explicit, widespread repression of critics and political opponents also added meaning to prompt compliance. Embracing the slogan "We can do it" meant that you never avoided challenges. If you could not do it, the reason could be that you were not doing it properly. Also, any failure to meet demanding completion dates and budgets risked giving the appearance of political resistance. When a goal was set and plan put in place, the focus shifted quickly to minimizing the time to achieve it, and at minimal cost. Setting a goal also fixed the idea that achieving it in the most effective manner was possible. Those responsible for its realization typically felt pressure to go all out, day and night, to reach the goal as soon as possible. The earlier, rapid construction of the Gyeong-bu Expressway became a model for all to emulate.

This logic carried over into private realms, including government-set goals for personal saving, literacy, family planning, foods, and consumer behavior. It was not unusual, for example, for children to have their lunch boxes checked to make sure their families were embracing new mixed grains, and not relying entirely on rice.

To the same extent that quantitative economic goals attracted focused, persistent attention, so was there official negligence of areas that were not easily quantifiable nor associated with the program of industrialization for export growth. These included the living conditions of those who did not contribute to or benefit from the program.

KOREAN MIRACLE? CONTINUING STRUGGLES IN THE SECOND ECONOMY

Despite all these initiatives, President Park never succeeded in scaling up to dominance his nationalist images of Koreans elevating themselves and their country through export industries. Park's vision never seriously challenged the dominant image of the humanist scholar-official who warranted his elite status and powerful position by virtue of advanced literary understanding and moral development. Throughout Park's eighteen-year rule, the interest of families in preparing their children for academic entrance examinations to the maximum extent possible not only continued unabated but actually intensified. For families competing to get their children into top elementary

---

[30] O, *Park Chung-Hee-Neun Eotteogge Gyeongjeganggug Mandeureonna* (*President Park Chung-Hee's Leadership and the Korean Industrial Revolution*), 2010, 21.

schools, middle schools, high schools, and universities, technical work still carried associations with the hands alone, to the exclusion of the mind and spirit. Disciplining the individual spirit led most directly to images of service to the whole. Park advanced his program "in the face of a chronic legitimation crisis."[31]

Beyond the country's borders, the period of Park's rule became known as the "Korean miracle." The inequities alone generated during the period make the term debatable, if not wholly inappropriate. For our purposes, even for technical personnel created with industrialization and, hence, engineering, a primary outcome was ambivalence. The Park government never fully matched its development agenda, dedicated to increased industrial production and exports, to an educational agenda championing training and education in vocational and technical areas.[32] It was reasonably successful for men at the lowest and highest statuses, but much less so for those in the middle. And its contributions to the technical training and education of women were minimal, typically aimed at preparing them for the lowest-status labor.

For the lower-level technical workers, including women, the Park government did increase employment to unprecedented levels. During the 1960s, employment opportunities exploded in the labor-intensive light industries. Official unemployment fell from 20% to single digits by the end of the decade.[33] During the 1970s, the semi-skilled workers who attended the targeted secondary schools were typically able to find jobs in those industries for which they had prepared themselves. Even semi-skilled workers earned extremely low wage rates, however, and those higher-level pathways were essentially unavailable to women.[34]

The Park government successfully called into workplaces young men and women from rural areas to the cities, young boys who were poor but smart into technical schools, and unmarried girls and women from patriarchal families into low-level schools and work. It celebrated them as "industrial workers," "frontrunners of export," or, best, "flag-bearers of national modernization."[35] Many new workers firmly believed that devoting themselves to the new work would secure stable lives for their families and advance their country. The so-called "spirit of hungry" challenged all directly.

During the 1960s and 1970s, the average family size exceeded five members, with 25% of families including three generations. Many families pooled and concentrated their resources to educate the most promising child, usually a son. Others, especially daughters, found themselves charged to provide support. Most transported workers successfully endured the harsh conditions of

[31] Lie, *Han Unbound*, 1998, 112.

[32] Kim, *Hanguk-Ui Gongeopjeongchaek Yeongu* (*Research on the Industrial Education Policy of Korea*), 2001, 123; Seth, *Education Fever*, 2002, 5.

[33] You, *Gyeongje Seongjang Sinwha-Ui Heowa Sil* (*The Truth and Falsity of Economic Growth*), 2011, 63.

[34] Lie, *Han Unbound*, 1998, 104–108.

[35] O, *Park Chung-Hee-Neun Eotteogge Gyeongjegangggug Mandeureonna* (*President Park Chung-Hee's Leadership and the Korean Industrial Revolution*), 2010, 445–447.

long hours and low wages. Yet workers in factories never overcame the denigrating labels of *gong-suni*, factory girl, and *gong-dori*, factory boy.

On the high end, President Park became the ultimate patron for mostly male scientist-engineers who completed higher education, especially abroad but at home as well. He offered scientist-engineers the prospect of becoming national figures with national identities, icons of industrialization. For the Ministry of Education, official prestige for positions in the technical bureaucracy "resulted in less resistance to higher quotas for these programs."[36] In contrast with engineers from earlier eras who received little external affirmation and found themselves forced to pursue individual pathways for individual reasons, the Park government offered both high status and a sense of connectedness to the country as a whole.

Official recognition also produced high enrollments. One significant outcome during the 1970s was a "glut in highly trained scientists and engineers" on the job. Many scientist-engineers competed for an insufficient number of positions, at least until the growth of private conglomerates during the 1980s ultimately provided lucrative positions with upward career mobility.[37] Furthermore, despite attracting high numbers, new technical curricula in engineering and the sciences that could lead to positions as scientist-engineers typically remained lower in status than existing curricula in the humanities and social sciences, especially law, philosophy, and economics.

In addition, enrollments at the mid-level vocational junior colleges "fell far short of that planned."[38] The government responded by shifting scholarship support rather dramatically in their direction. Although nearly 40% of all state scholarships went to them in 1970, vocational college graduates still carried the stigma of technical work without the promise of pathways to prestigious positions.

Finally, Korean businesses were themselves "not very enthusiastic" about providing training centers in-house. They too "showed a preference in hiring non-vocational graduates, feeling that technical skills [but not moral education and character] could be picked up on the job."[39]

Acutely aware of continuing popular resistance to his plan to build collective national identity through industrialization and exports, President Park attempted in a 1974 press conference to reframe higher education in terms of "mental nationality." "The fundamental purpose of college education," he asserted, "is to educate competent talents who can be actively devoting themselves to the development of the nation and national restoration." To that end, he continued, "[W]e should not deliver college education which raises global citizens without any mental nationality." People needed to reorient themselves. "[W]e should educate sound Koreans," he said, "who are equipped

---

[36] Seth, *Education Fever*, 2002, 128.

[37] Seth, *Education Fever*, 2002, 128.

[38] Seth, *Education Fever*, 2002, 127.

[39] Kim, *Hanguk Gyo-Yukui Sahoehakjeok Jindangwa Cheobang (Sociological Diagnosis and Prescription for Korean Education)*, 1998, 90; Seth, *Education Fever*, 2002, 127–128.

with thorough historical views with 'education with nationality'. ... This is the purpose and mission of education."[40]

The President energetically promoted activities in the second economy, the economy of spirit, to help scale up his vision of techno-educational progress through industrialization. A few months after establishing the Ministry of Science and Technology in 1967, for example, Park founded the Korean Association of Science and Technology Supporters. One objective was to help provide financial support for elder scientist-engineers. But as its name indicates, its larger goal was to build a community of advocates for science-technology and, ultimately, industrial expansion.

Park also worked to scale up an image of his new Korea as a resurrection and reconstruction of an old Korea rather than something produced in response to EuroAmericans. In 1969, his government established a commemorative society for Jang Yeong-sil (1418–1450). By many accounts, Jang Yeong-sil was the best technician of the Joseon Dynasty. He invented a sundial, an iron printing technique, and a device for celestial observations. Although Jang Yeong-sil was born a slave, King Sejong (1397–1450) elevated him to the status of a government bureaucrat, an early scholar-official. Promoting the image of upward mobility through dedicated technical work, the new commemorative society established the Jang Yeong-sil Science and Culture Award for the best *gwa-hak-gi-sul-ja*, either at home or abroad.[41]

Beginning in 1970, the New Village Movement, *Sae-ma-eul-undong*, had been providing construction materials to rural areas to support new infrastructures and attract villagers into the country-wide project of industrial development. The Movement to Promote Science to the Whole Nation, *Jeon-gungminui-gwahak-hwa-undong*, initiated in 1972, called on all citizens to learn at least one skill or technique so they could both improve their own lives and contribute to the country's development.[42] "What is promoting science to the whole nation?" President Park asked provocatively in a 1973 speech. "It is making people's thinking and living more scientific so that they know how to utilize even a simple and basic scientific knowledge efficiently, in the New Village Movement as well as in forestation and reforestation projects."[43] Yet the mere fact that the President had to keep making such speeches and promote his image of technological and personal advancement through industry proved instructive. These attempts to persuade provided telling evidence that he was not fully capturing his audience.

[40] The Presidential Secretariat, *Park Chung-Hee Daetonglyeong Yeonseolmunjip 5* (*Speech 5 by President Park Chung-Hee*), 1976, 238–239.

[41] Song, "Joseon Sidae Choego-Ui Gisulja Jang Yeong-Sil (Jang Yeong-Sil, the Best Scientist in the Joseon Period)," 2008, 29.

[42] Moon, "Park Chung-Hee Sidae Damhwamuneul Tonghae Bon Gwahakgisuljeongcheakui Jeongae (A Discourse Analysis of Science and Technology During the Park Chung-Hee Era)," 2012, 101.

[43] The Presidential Secretariat, *Park Chung-Hee Daetonglyeong Yeonseolmunjip 10* (*Speech by President Park Chung-Hee 10*), 1973, 113.

President Park's technocrats, and those who aspired to those positions, appeared to be wholly on board with his mission to scale up a collective national identity through industrialization. And why not? Such provided pathways to enhanced status. As we see in the next chapter, their association with this vision also continued after it ended. In this case, however, the commitment marginalized them when a different vision of the country scaled up to dominance. As we will also see, many joined those who protested the repressions of the Park regime and embraced new images of democratic citizenship.

REFERENCES

Cho, Hwang-hee, Author and Author. *Hanguk-Ui Gwahakgisul Illyeok Jeongchaek* (*Review of Science & Technology Human Resource Policies in Korea*). Seoul: STEPI (Science and Technology Policy Institute), 2002. 89, 90

Choi, Hyeongseop. *Bul-I Kkeojiji Anhneun Yeonguso* (*The Research Institute: Its Light Never Goes out*). Chosun Ilbo Co., 1995. 86, 93

Hahn, Young-Whan. "Administrative Capability for Economic Development: The Korean Experience." <manuscript>, 1995. 79

History, National Institute of Korean. *Geunhyeondae Gwahaggisulgwa Salmui Byeonhwa* (*Science, Technology, and Changing Life in Modern Times*). Seoul: Doosan Dong-A, 2005. 84

Hong, Ha-sang. *Lee Byung-Chul Dae Chung Ju-Yung* (*Lee Byung-Chul Vs. Chung Ju-Yung*). Seoul: Hankyung BP, 2011. 81

Hughes, H. "Why Have East Asian Countries Led Economic Development?" *Economic Record* 71, no. 212 (1995): 88-104. DOI: 10.1111/j.1475-4932.1995.tb01874.x. 79

Jung, Yong-uk. "5.16 Kudeta Ihu Jisikin-Ui Bunhwa-Wa Jaepyeon (The Differentiation and Reorganization of Intellectuals after the 5.16 Coup)." In *1960 Nyeondae Hangug-Ui Geondaehwa-Wa Jisikin* (*Korean Modernization and Intellectuals in the 1960s*), edited by Jung, Yong-uk, Seoul: Sunin, 2004. 92

Kim, Deok-gi. *Gukgagisuljagyeok Jedo-Ui Hyeonhwang-Gwa Gaeseon Bangan* (*Current Status and Recent Issues in the National Technical Qualification System*). Seoul: KRIVET (Korea Research Institute for Vocational Education & Training), 2004. 90

Kim, Dong-hwan. *Hanguk-Ui Gongeopjeongchaek Yeongu* (*Research on the Industrial Education Policy of Korea*). Seoul: Moonum, 2001. 90, 95

Kim, Geun-bae. "Gwahakgisul Ipgukui Haebudo: 1960 Nyeondae Gwahakgisul Jihyeongdo (Anatomical Chart of a Scientific and Technological State: The Topography of Science and Technology in South Korea in the 1960s)." *Yuksa Bipyeong* 85, no. (2008): 236-261. 87, 91

Kim, Hyun-mee. "Hangugui Geundaeseong-Hwa Yeoseong-Ui Nodong Gwon (Modernity and Women's Labor Rights in South Korea)." *Journal of Korean Women's Studies* 16, no. 1 (2000): 37-64. 90

Kim, Kyung-dong. *Hanguk Gyo-Yukui Sahoehakjeok Jindangwa Cheobang* (*Sociological Diagnosis and Prescription for Korean Education*). Seoul: Jipmoon, 1998. 96

Kim, Lin-su. "Gwahakgisul Jinheung-Gwa Gyeongje Baljeon (The Promotion of Science, Technology, and Economic Development)." In *Hanguk Geundaehwa, Gijeokui Gwajeong* (*Modernization of the Republic of Korea: A Mirachulous Achievement*), edited by Cho, I-je and Carter Eckert, Seoul: Monthly Chosun, 2005. 87

Kim, Linsu. *Imitation to Innovation: The Dynamics of Korea's Technological Learning*. Boston: Harvard Business School Press, 1997. DOI: 10.1057/jibs.1997.42. 85

Kuznets, Paul W. "Government and Economic Strategy in Contemporary South Korea." *Pacific Affairs* 58, no. 1(Spring) (1985): 44-67. DOI: 10.2307/2758009. 79

Lee, Deok-jae. "Park Chung-Hee Jeongbu-Ui Gyeongje Jeongchaek (Economic Policies in the Park Chung-Hee Era: Political Economy of a Double-Edged Sword)." *Yoksa wa Hyonsil*, no. 74 (2009): 79-112. 83

Lee, Eun-kyung, "Gwahak Gisulgwa Yeoseong-Ui Jeongchaek Jaeng Jeom (Policies at Issue for Science-Technology and Women)," Seoul: STEPI, 2001. 90

Lee, Ho. *Sindeullin Saramdeul-Ui Habchang* (*Chorus of the Bewitched*). Seoul: Hansong, 1998. 88

Lie, John. *Han Unbound: The Political Economy of South Korea*. Stanford, CA: Stanford University Press, 1998. 82, 95

Ministry of Education. *Tonggyero Bon Hanguk Kyo-Uk-Ui Baljachwi* (*The Evolution of Korean Education as Seen through Its Statistics*). Seoul: MOE, 1997. 91

Ministry of Education (MOE), "Gyoyuk Tonggye Yeonbo (Statistical Yearbook of Education)," Seoul: MOE, 2000. 90

Moon, Man-yong. "Park Chung-Hee Sidae Damhwamuneul Tonghae Bon Gwahakgisuljeong-cheakui Jeongae (A Discourse Analysis of Science and Technology During the Park Chung-Hee Era)." *Journal of the Korean History of Science Society* 34, no. 1 (2012): 75-108. 91, 97

O, Won-chul. *Park Chung-Hee-Neun Eotteogge Gyeongjeganggug Mandeureonna* (*President Park Chung-Hee's Leadership and the Korean Industrial Revolution*). Dongsuhbook, 2010. 77, 90, 92, 94, 95

Pohang Iron and Steel Company (POSCO). *Posko 35 Nyeonsa (35-Year History of Posco)*. Seoul: POSCO, 2004. 84

Seo, Kap-kyoung. *Chulgangwang Park Tae-Jun Yiyagi (Steel King: The Story of T. J. Park)*. Seoul: HanEon, 2011 [1997]. 87

Seth, Michael J. *Education Fever: Society, Politics, and the Pursuit of Schooling in South Korea*. Honolulu, HI: University of Hawai'i Press, 2002. 89, 90, 91, 96

Song, Sung-soo. "Joseon Sidae Choego-Ui Gisulja Jang Yeong-Sil (Jang Yeong-Sil, the Best Scientist in the Joseon Period)" *Korean Journal of Mechanical Engineering* 48, no. 11 (2008): 29-32. 97

The Presidential Secretariat. *Park Chung-Hee Daetonglyeong Yeonseolmunjip 10 (Speech 10 by President Park Chung-Hee)*. Seoul: The Presidential Secretariat, 1973. 97

The Presidential Secretariat. *Park Chung-Hee Daetonglyeong Yeonseolmunjip 5 (Speech 5 by President Park Chung-Hee)*. Seoul: Daehangonglonsa, 1976. 97

You, Jong-il, ed. *Gyeongje Seongjang Sinwha-Ui Heowa Sil (The Truth and Falsity of Economic Growth)*. Seoul: Sisa IN Book, 2011. 87, 95

CHAPTER 5

Loss of Privilege and Visibility: 1980–1998

In 1993, the chairman of Samsung Group's executive office published a training manual for senior officers. It was designed to help them cope with what it called "[c]hange in the values of the new generation. ..." The term "new generation" had become a widely used label for young, primarily male, university graduates in technical fields who resisted the top-down, military-style organizational practices that had peaked during the 1970s push for heavy and chemical industries. Its members, asserted the manual, "avoid self-sacrifice for the nation or others." They had developed "a way of life centered on themselves and their families." They had become largely "indifferen[t] to ... politics and society." The report went on to assert that the new generation, so different as to constitute a new race of Koreans, had a "strong tendency to be complacent" as well as "a weakened desire for promotion." While they did "tend to devote themselves to what they like" and even "find [the work] fruitful," they bore no affection for their employers, the big companies. Their commitment was "for the 'work' itself."[1]

President Park was turning in his grave. Fewer than 15 years had passed since his assassination at the hands of the director of the Korean Central Intelligence Agency. Park's program to elevate technical workers, especially engineers, to the status of scholar-officials, icons of his image of "mental nationality," had been supplanted by other initiatives. New images of Korea and Korean advancement relative to other countries had scaled up to dominance. During the 1980s, there was democratization and a shift in the leadership of industrial development from government to the *chaebols*. By the early 1990s, a new image of economic competitiveness had scaled up to replace the anti-communism of Park's era. Linked to democratization, the lens of economic competitiveness carried a novel techno-educational vision: it highlighted individual ingenuity and innovation over collective responsibility. As the government withdrew from the center of industrial development, the longtime focus on heavy and chemical industries gave way to more diverse initiatives in electronics, information, and communication technologies, and the production of people to fit them.

As Park's programs receded into the past, the official work of elevating engineering and the technical workers who contributed to it lost energy. From scientist-engineers to technicians, male practitioners of engineering lost coherence and visibility as official techno-national icons of the country's advancement. They spread out in diverse individual struggles to find stable work. In this environment, participation by women technical workers increased significantly.

---

[1] Cho, "Wiro-Buteo-ui Gieop-Hyeok-myeong (Corporate Revolution from the Top)," 1993, 79.

In the early 1990s, a movement of men engineering educators sought to reinvigorate a strong sense of service among men and women engineers in a new sense, as individual professionals. While they successfully established professional organizations, their efforts to recapture national visibility for engineers per se, apart from the more general category of scientist-engineers, struggled to gained traction. The East Asian financial crisis of 1997 further inhibited these efforts. Employment became uncertain in every sector and the "new generation" of graduate engineers increasingly pursued individual careers, self-development, and entrepreneurship in ways that no longer attracted the national spotlight.

RATIONALIZING THE ECONOMY

It was clear across Korea by the early 1980s that the military regime of President Chun Doo-Hwan (b. 1931) (Figure 5.1) would not be able to keep dominant the practices of export-led growth and political repression that had come to define the Park regime. Apart from not exhibiting the personal charisma and visionary passion of President Park, President Chun had to confront dramatically different sets of circumstances, both locally and internationally.

Figure 5.1: President Chun Doo-hwan. Source: National Archives of Korea.

The Park economic strategy of attempting to pick and support winners in industry with the assistance of technically trained leaders lost its tacit acceptance by government officials. When energy prices rose dramatically in 1979, the year of Park's death, and then the country experienced negative economic growth in 1980 for the first time, government economists and economic advisers successfully made the case that the Park economic system was inefficient, even irrational. Chun's new government created and sought to scale up a new vision of the Korean economy led by private-sector initiatives and decision-making rather than by government.[2]

At first glance, the Chun government was participating in a worldwide withdrawal of governments from stimulating and directly participating in local commercial economies (but, notably, not military economies). Margaret Thatcher had become prime minister of the United Kingdom in 1979 and Ronald Reagan became the U.S. president in 1981. What became known as Thatcherism and Reaganomics together involved, among other things, reducing direct government involvement in regulating commercial enterprises as well as privileging those capable of investing in those enterprises. In the U.S. and Europe, the term "neoliberalism" emerged to name the many forms and outcomes of this process, including concentrating capital and economic power in fewer and fewer hands outside of government.

The Chun government made Korea stand out by continuing to fund private economic initiatives. Yet it broadened and diffused the focus of governmental action from selecting specific private industries it would support and control to funding infrastructural support and basic technology that could stimulate a wide range of industries.[3] Even this change, however, had the effect of shifting economic authority and power away from the public and into private hands.

On the ground, rationalizing the economy meant (a) minimizing overlaps in production among *chaebols*, (b) reducing the number of target industries that received direct government support, and (c) rectifying what government economists argued was an imbalance among industries in favor of the heavy and chemical industries. The military Chun government initially added practices of economic rationalization to the still-dominant image of national security, protecting Korea from external threat.

In 1980, its Special Committee for National Security Measures added economic directives to its security policies, for example. The Committee scaled back government investments in automotives, power plants, heavy equipment for construction, electrical equipment, diesel engines, and electronic switching systems. It also reduced overlaps in the fertilizer, marine, and coal mining industries, and began liquidating government-owned companies. During a two-year period

---

[2] Park, "Jeongbu Gi-Eop Gwangye-Ui Dayang-Seongkwa Geu Gyeoljeong Yo-in (Government-Business Relations in Korea)," 2000, 571.

[3] This strategy paralleled that adopted in Japan during the 1970s and peaking in the 1980s, led by the Ministry of International Trade and Industry. The literature on the Japanese initiatives is voluminous, especially after U.S. government agencies began attempting in the 1980s to emulate what they thought was taking place in Japan. A good starting point is the work of Tessa Morris-Suzuki. See Morris-Suzuki, *A History of Japanese Economic Thought*, 1989; Morris-Suzuki and Seiyama, *Japanese Capitalism since 1945*, 1989; Morris-Suzuki, *The Technological Transformation of Japan*, 1994; Morris-Suzuki, *Re-Inventing Japan*, 1998.

later in the decade, for example, the Chun government ordered the transfer of 56 enterprises to private firms.[4]

A telling example of the new circumstances facing the Chun vision came in a 1980 confrontation with the Hyundai *chaebol* group. Even though the Hyundai Motor Company had been in operation since 1967, the Security Committee sought to limit automobile production to the Daewoo group. Members of the Security Committee expected the Hyundai leadership to agree to target its heavy industry work on power plant construction. Power plant projects came with guaranteed government contracts, while the automobile industry had uncertain markets.

Much to the Committee's surprise, Chairman Chung Ju-yung insisted on entering the auto industry because it promised more stable levels of production and operation in the future (Figure 5.2). Power plants depended on individual orders, and the economy at the time was experiencing negative growth. "If we gave up automotives," Chairman Chung later said, "we could never do it again."[5] By relenting and dropping its directive, the Security Committee effectively acknowledged and affirmed a power shift toward the *chaebols*.

Figure 5.2: Ulsan plant view (left) and assembly line (right) in the 1980s. Source: Hyundai Motor Company.

---

[4] Kuk, "80 Nyeondae Busil Gi-Eop Jeongni Gwajeong-Eul Tonghaebon Guk-Gawa Daegi-Eopui Gwangye (The Relationship between the Government and Chaebols through the Liquidation Process of Ailing Companies in the 1980s)," 1990, 146.

[5] Lee, *Gyeongje-Neun Dangshini Daetongnyeong-Iya* (*You Are the President in the Economy*), 2008, 126–128.

The Chun government developed a multitude of policies to provide infrastructural support for a wide range of industries. The administration's National R&D Project (NRP) in 1982 and Industrial Generic Technology Development Project (IGTDP) of 1987 were two prominent examples, shifting away from the heavy and chemical industries. While IGTDP concentrated mainly on current problems in existing technology areas with high economic prospects, NRP focused primarily on future problems in new technology areas that offered great promise and carried a high risk of failure.[6] Formal legislation announced wholly new initiatives. The Industrial Development Act of 1986 provided investment support for technological development and productivity improvement according to function rather than by industry.[7]

The Technology Development Promotion Act of 1981 provided support to private companies to establish research institutes. The Act charged these private R&D organizations to develop technologies that would benefit a wide range of industries. It also gave them the flexibility to consider the medium- to long-term in calculating economic returns.

The new "enterprise institutes" would supplement, and later supplant, the work of government-supported research institutes (GSRIs) that had focused on specific industries. The government reduced the 16 GSRIs founded during the 1970s down to nine, redirecting them to implementing specifically national R&D projects. For example, in collaboration with private enterprise institutes, GSRIs developed such technologies as digital electronic switching and electronic protocols.[8] The Code Division Multiple Access (CDMA) project proved especially important. Coordinated by the Electronic Telecommunications Research Institute (Figure 5.3), and with participation by Samsung, LG, Daewoo, and Korea Telecommunications, CDMA provided a key technology for cell phone communication. It enabled several radio transmitters to send signals over a common communication channel.

---

[6] Kim, *Imitation to Innovation*, 1997, 50–51.

[7] Yun, *Gwahak Gi-Sul-Gwa Hanguk Sahoe (Science-Technology and Korean Society)*, 2000, 159.

[8] Ministry of Education, "Gwahaggisul 40 Nyeonsa (Forty-Year History of Korean Science and Technology)," 2008, 115.

Figure 5.3: Electronic Telecommunications Research Institute (1995). Source: ETRI 35-Year History.

Promoting a wide range of industries did not mean, however, shifting power away from the conglomerates to small- and medium-sized enterprises. Despite its name, the Monopoly Regulation and Fair Trade Act of 1981 in fact authorized *chaebols* to expand their enterprises into new industries. During the five-year period beginning in 1984, the proportion of industrial markets occupied by the 30 largest conglomerates increased from 63% to 75%. And during the fourteen-year period beginning with the new law, the government issued a total of only 62 corrective orders.[9] In exchange for this freedom to expand, the *chaebols* were expected to place greater public emphasis on their responsibilities to customers and to avoid trading practices that might lead to monopolistic control.

Overall, the Chun government worked hard to scale down Park's vision of "a government-led protective economic operation" and scale up an image of "private-led market competitive and open-style economic operations."[10] As a guiding slogan, it promoted "Conversion to a Private-sector-led Economy."[11]

In response to this withdrawal by government, the *chaebols* elaborated their own policies and practices of technological development. By the mid-1980s, many had encountered significant difficulties in licensing new technologies from U.S. and European firms. The change was due in part

---

[9]   Jang, "Hangugui Gaebal Guk-Ga (Korean Developmental State)," 1999, 97.

[10] Bae, "Gwonwiju-Eu Jeongchi Chejeha-Eu Jeongbuwa Gyeongje Iik Jibdan Gwangye (Relationships between Governments and Economic Interest Groups in Authoritarian Political Regimes)," 2001, 31.

[11] Park, "Jeongbu Gi-Eop Gwangye-Ui Dayang-Seongkwa Geu Gyeoljeong Yo-in (Government-Business Relations in Korea)," 2000, 571.

to shifting images of international relations and international competition taking place in those countries as well. The U.S. military had clearly won its nuclear arms race with the Soviet Union by the mid-1980s. The Soviet Union could not keep up with U.S. allocations to a non-productive arena that provided no economic return. Less than half the size of the U.S. economy, the Soviet economy had been relatively stagnant since the 1970s. In the 1980s, Soviet leader Mikhail Gorbachev initiated a process of political liberalization (one version of neoliberalism), leading ultimately to the Soviet Union's dissolution in 1991.

Meanwhile, across the U.S. a new image of international competition was scaling up as early as 1983. In place of a grand confrontation between capitalism and communism, the defining feature of the Cold War, this new vision pictured the planet as a collection of countries competing with one another for economic dominance. As threats from the Soviet Union and communism appeared to decline, the soon-to-be dominant image of "economic competitiveness" identified Japan and East Asia, including Korea, as posing the greatest risks to U.S. well-being. They threatened to infuse the U.S. economy with consumer products of increasing quality. U.S.-based companies, and increasingly European companies (with new identities added by the European Union on similar grounds), found themselves re-imagining East Asian conglomerates as genuine competitors. Freely sharing technologies through licenses became potential threats to national security, and the flows of technology transfer sharply declined. When seen through the techno-political[12] lens of economic competitiveness, Japan and Korea posed threats to the U.S. despite the fact that the U.S. military maintained tens of thousands of military personnel in both countries. Such was the force of the new image.[13]

Many Korean companies responded by launching efforts to develop new technologies. The electronics firm LG (once Lucky Goldstar) successfully developed a technology for color television (Figure 5.4). The Hyundai Motor Company concentrated on engine development. Through most of the 1980s, Hyundai had obtained engines from the Japanese company Mitsubishi. After more than four years of work and nearly five million dollars, Hyundai independently developed the alpha engine, its first engine for compact cars, in 1991.[14]

[12] Hecht, *The Radiance of France*, 1998.

[13] One can make the case that the new image of economic competitiveness in the U.S. appropriated, adapted, and localized images of economic competition that had previously scaled up dominance across Japan and Korea. The governments of both countries had long identified economic competition via exports as vehicles for national progress and parity. In the 1980s, the U.S. government even attempted experiments in government-industry-university collaborations (Downey, "The World of Industry-University-Government," 1995; Downey, *The Machine in Me*, 1998.). What ultimately scaled up across the U.S. and came back, as described above, was the neo-liberalist stipulation that governments reduce their involvement in directing or regulating private economic activity at home. What also came back were new barriers to trade.

[14] Kim and IMI, "Hyundai Motor Co.," 2007, 107.

Figure 5.4: Lucky Goldstar in 1970. Source: National Archives of Korea.

ENGINEERS LOSE THE SPOTLIGHT: 1980s

By shifting the locus of economic development away from the government and into the private realm, the Chun policies and practices had the cumulative and ironic effect of simultaneously multiplying engineers and marginalizing them. On the one side, expanding R&D activities in private firms dramatically increased demands for technical workers who were university graduates, especially the *gwa-hak-gi-sul-ja*. The Hyundai Motor Company increased the number of R&D engineers it employed, for example, from under 200 in 1975 to more than 3,800 in 1994. Half of those had earned a master's degree or higher.[15] And to accommodate rapid increases in demand, the education ministry sharply increased quotas for admission to engineering majors at universities. During the nine-year period, 1980–1989, the number of engineering graduates tripled, from roughly 10,000 to more than 30,000.

Yet even as the *chaebols* rapidly expanded the scope of their operations, small and medium-sized firms in the manufacturing and service sectors grew and multiplied even faster. In 1982, roughly 45% of technical workers served in companies with more than 300 workers. By 1987, that proportion had dropped to 39%, and to 34% by 1993. The rapidly expanding computer and elec-

---

[15] Kim, *Mobang-Eseo Hyeokshin-Euro (Imitation to Innovation)*, 2000, 159.

tronics industries contributed to this trend. Smaller companies found it easier to locate niches for themselves in those markets.

The trend was also due to another massive transformation taking place across the country. With the threat of communism scaling down as a defining Korean reality, the Chun government faced rapidly growing resistance to political repression. How could the government justify repressing dissent at home when it could no longer point to a nearby evil enemy posing a seemingly clear military threat? The most prominent political stories of Korea during the 1980s and 1990s document expanding dissent and resulting democratization (Figure 5.5).

Figure 5.5: Democratization movement in July 1987. Source: Korea Democracy Foundation and Kyunghyang Shinmun.

University students provided a primary source of resistance. By virtue of the pathways they were following to become scholar-officials, university students could claim the right to name and resist what they considered to be the failures of leadership. Indeed, the expansion of higher education for engineers was actually a small part of a massive expansion in admission to higher education more generally. Enrollments swelled from roughly 210,000 in 1975 to more than a million by 1985. With no grounds to resist expanded demand for higher education, facilitating its expansion as quickly as possible became official Chun policy.[16]

---

[16] Cho et al., *Hangug-Ui Gwahakgisul Il-Lyeok Jeongchaek* (*Review of Science and Technology Human Resource Policies in Korea*), 2002, 105.

Workplaces also became primary sites of resistance. President Park's Economy First principle had justified low wages in the name of expanded exports. Under Chun, this justification disappeared. Labor unions formed and carried out strikes for increased wages and other benefits (Figure 5.6). In 1987 alone, for example, the number of labor disputes during the decade exceeded 3,700.[17]

Figure 5.6: Labor strike at Ulsan Industrial District in 1987. Source: Hyeonjang Silcheon Labor Alliance.

Political dissent might be described as the nomination of new images of governance, coupled with efforts to scale them up. During the 1980s, there was great excitement across the country about ending military control and building a society and government that might warrant the name "democratic." Intellectuals debated what might constitute a valuable and desirable life, in the hope of guiding future regimes. Plans to host the Asian Games in 1986 and summer Olympics in 1988 added urgency, for international media would soon turn to explain Korea to the world (Figure 5.7).

[17] Kim, "Urinara Nosa-Gwangye-Ui Sidaebyeol Byeonhwa-E Gwanhan Yeongu (History of Korea Labor-Management Relations from the 1940s to the 2000s)," 2009, 295.

Figure 5.7: Official mascot of the Seoul 1988 Olympic Games. Source: Seoul Olympic Museum.

President Chun's successor, Roh Tae-woo (b. 1932) (Figure 5.8), attempted to frame and promote the economic policies of his government as serving "ordinary people." Roh had been Chun's classmate at the Korea Military Academy and became a senior official in his administration. With public resistance continuing to grow in scale and intensity, Chun named Roh to lead their party. Roh supported amending the constitution to permit direct popular election of the next president for a single five-year term, and then became that next president in 1987. To highlight ordinary people in his administration effectively meant that "nothing could be discussed without mentioning equity and welfare."[18] Yet governance by Roh was still military governance.

---

[18] Lee, *Sillok Yuk Gong Gyeongje* (*A True Story*), 1995, 41.

Figure 5.8: President Roh Tae-woo. Source: National Archives of Korea.

The collective, persistent work of scaling up political democracy actually culminated with the 1993 election of Kim Young-sam (b. 1927) and, hence, a civilian government. Democratization redirected the spotlight away from engineers and heavy industry. In shifting the locus of control over technological developments to the private sector, President Chun had relied more on economists than on scientist-engineers. The senior secretaries to the president for economic affairs, for example, Sa Gong-il and Kim Jae-ik, had both studied economics in the U.S. As scholar-officials, these economists "had a belief in the importance of market and competition, and preferred development strategies based on the creativity of private sectors and competitive advantage." Dr. Kim actually "taught President Chun economic theories in an informal setting, . . . persuad[ing] the President to support the principle of stabilization in all economic policies."[19]

Chun had also reduced from 38% to 15% the proportion of economic advisers who were on staff as fulltime bureaucrats. External experts thus became the new thing. As such, they had little chance of gaining the direct administrative authority that President Park had earlier granted to the *gwa-hak-gi-sul-ja*.[20]

Declining access to the dominant image and practice of scholar-officials proved decisive for graduates of engineering degree programs. The numbers of engineering graduates working in the private sector increased dramatically. Attending university in engineering certainly continued to provide pathways to stable, secure positions in research and development in the *chaebols*. But within large, private organizations, graduate engineers had no particular grounds to claim distinct status

[19] Kim, *Young-Yogui Hanguk Gyeongje* (*The Glory and Shame of the Korean Economy*), 1999, 267.

[20] Chung, "Daetongnyeong-Ui Uisa Gyeol-Jeong-Gwa Jeonmun Gwallyo-Ui Yeokal (Presidential Decision Making and the Role of the Bureaucrat)," 1989, 78.

from other technical experts. Their education emphasized expertise in the engineering sciences via classroom lectures, textbooks, and semester exams. Although their dominant subsequent identities were as employees of the large companies, those companies did not (yet) seek significant influence in the contents of engineering curricula.[21] Control remained largely in the hands of faculty who either focused on teaching or began to emulate engineering faculty from the U.S., Europe, and Japan by beginning to undertake research. In any case, for graduates of engineering degree programs, engineering science problem solving to advance a company brought nowhere the visibility of working to build industry for the country as a whole. Techno-corporate formation did not necessarily lead to techno-national identities.

Furthermore, work in small and medium-sized enterprises did not necessarily promise leadership and high status either. The 1980s labor disputes had started with production workers but had spread to office workers and professional workers as well. In a favorable economic environment that included low oil prices, low interest rates, and low exchange rates (the "three lows"), the large conglomerates sought to mollify workers by increasing wages. Increases amounted to roughly 10% in 1987 and then 21% in each of the following two years. The small and medium-sized enterprises, however, were not able to follow suit. Working there actually fell in attractiveness relative to the large conglomerates.

During the 1970s, anyone who worked technical jobs in heavy industry had been able to claim they did engineering, and Park Chung-hee's government defined doing engineering as advancing Korea. Technical workers working on the many new, small-scale electronics technologies during the 1980s would not have access to such official endorsement. Elite *gwa-hak-gi-sul-ja* still had stability and jobs, especially in *chaebols*, but even they found themselves increasingly distanced from the main centers of power decision making. Official efforts to build connections between industry and universities produced "cacophony" as much as coordination.[22] They were disappearing inside the powerful private-sector enterprises.

COORDINATED CREATIVITY?

It was a new spring in 1993, or so the president told everyone. The death of President Park in 1979 had stimulated what had become known then as the Spring of Seoul. In retrospect, however, that so-called spring had yielded more than a decade of continued military rule. The 1993 election of a civilian government overlapped with recovery from a recession. The term became usable again. "In the past," asserted the civilian president Kim Young-sam (Figure 5.9) at his inauguration, "our

---

[21] Cho, "Saneop-Gye Yogu Insik Suyonggineung Ganghwahaeya (Need to Strengthen the Acceptance of Industrial Demand)," 1992, 31.

[22] Park, "Sanhak Hyeob-Dong-Gwa Saneop Deahak Gyouk Banghyang (Industry-University Cooperation and Education at Industrial Colleges)," 1991, 71.

nation had both a thriving summer and shrinking winter." Now a "new spring is coming to our nation," a spring that would require "new commitments and new starts."[23]

Figure 5.9: President Kim Young-sam. Source: National Archives of Korea.

President Kim Young-sam was attempting to return to an image of a unified Korea. Calling for coordination once again at a national level, his plans actually bore some resemblance to those implemented by President Park two decades earlier. But Park's Korea had been committed to the single-minded pursuit of increased exports, led by the government and enacted by technical workers. In President Kim's vision, the new Korea highlighted diverse contributors with variable ends. Unlike the military governments of the past, Kim's government would attempt to facilitate and support a range of contributors without requiring or demanding quasi-military discipline from them.

Kim's image of what he called the "new economy" emphasized creativity. Scaling up this image depended upon successfully characterizing the recent past as a period of decline from the economic heydays of the Park regime, a decline whose causes were internal. The "diligence and creativity" that had been "envied by people all over the world" during that period, according to President Kim, had been "disappearing." He pointed not to "challenges from the outside" world but to a "mental defeatism spreading inside ourselves."[24] Mental defeatism had to be rejected and left behind in favor of creativity.

[23] The Presidential Secretariat, *Kim Young-Sam Daetonglyeong Yeonsiolmunjip 1* (*Speech by President Kim Young-Sam*), 1994, 56.

[24] The Presidential Secretariat, *Kim Young-Sam Daetonglyeong Yeonsiolmunjip 1* (*Speech by President Kim Young-Sam*), 1994, 56.

Like those of his predecessors, President Kim's image of creativity was about catch-up. Korea could not catch up with powerful Western countries, he contended, simply by providing public support for infrastructural development and technologies that might be usable by multiple industries, as the Chun and Roh governments had attempted to do during the 1980s and early 1990s. It rather had to return to energetically expanding the creative work of technology development. The Minister of Science and Technology had in 1990 called for "self-supported … technology innovation … [in] value-added industries and high-tech industries."[25] It was just that the government of a new Korea, a civilian-led government, could not directly produce technological creativity nor even demand it by official fiat. To avoid appearing reactionary, especially to countries now also focused on economic competitiveness, it had to coordinate and facilitate rather than control.

The Kim push for creative technological development emphasized coordinated action by universities and *chaebols*, adding new identities to both. In the first place, universities greatly expanded their activities in technological research and development. Prior to the 1990s, the primary mission of most universities had remained that of higher education. Initiatives to encourage university-based research had begun emerging in 1990, prior to the election. The Kim government latched onto these, granting them priority and rapidly accelerating their growth. President Kim, in particular, authorized the Presidential Advisory Committee on Education Reform and the National Science and Technology Commission to not only devise new approaches to R&D at universities but also present their recommendations directly to him. He clearly wanted to prevent any detractors from using the bureaucracy to dilute or undermine creative new ideas and practices.

Secondly, the industrial conglomerates gained even greater visibility and power. Remember that the *chaebols* dominated the private sector in Korea and were owned by powerful families. Having long been subordinate to military governments through regulation and direct control, many *chaebols* responded to deregulation by the civilian government by enthusiastically, even aggressively, embracing the policies of the "new economy."[26] Rather than owning local rural estates and controlling local political economies, powerful families of the 1990s controlled large industrial networks extending both across the country and beyond its borders. Increased autonomy for the *chaebols* effected a dramatic shift in power from the public to the private sectors.

With privilege did come responsibility. One effect of elevated visibility for *chaebol* managers was increased pressure to demonstrate high levels of responsibility to employees. In 1993, the Federation of Korean Industries published the Corporate Culture White Paper with this end in mind. Korean conglomerates did not originate the concept of corporate culture. The Federation was in fact attempting to transfer an image from U.S. corporations, in which the idea of a corporation having

[25] Chung, "Segye Hanminjok Gwahakgisulja Eaehoe Chisa (Korea Conference on Innovative Science and Technology Speech)," 1990.

[26] Kim et al., "Minjuhwa Ihu-Ui Hangukui Gug-Ga Sahoe Gwangye (State-Society Relations after the Korean Democratization)," 2006, 4.

a distinctive culture emphasizing individual improvement was already scaling up to dominance.[27] Korean conglomerates contrasted with U.S. corporations in that they had access to Confucian images of leadership, highlighting the responsibilities of leaders to those they led. Indeed, such images were difficult to avoid. Similar to the U.S. case, however, along with the idea that all employees in a given company were in it together came the concern that management was promoting the image of shared culture as a device to control employees.[28]

In response to the call for creative technological development, many *chaebols* further developed their own research institutes, even accelerating the process. They also gained considerable influence, both direct and indirect, over the contents of university-based research. Consider the hopes of one interested observer, offered during the year after Kim Young-sam's election:

> I am wondering if our universities can develop their own specialized and profound basic technologies so that companies can use them. Isn't it possible for each university to have world-class skills in its own unique area, so that the related basic technology can converge in it?[29]

These comments came from the leader of the LG group, Chairman Koo Cha-kyung (Figure 5.10). His question was not rhetorical. It fit a pattern of expanding connections between *chaebols* on the one side, with increasing autonomy and public visibility, and the public institutions of universities on the other, increasingly participating in research and development for private ends.

President Kim Young-sam's image of a new economy and new Korea with a civilian government, increased autonomy for conglomerates, expanded government coordination of research, and expanded research by universities also carried significant implications for the contents of higher education. The previously dominant image of anti-communism had helped to justify top-down control by military governments. The newly dominant image of a disaggregated Korea in economic competition with other countries put pressure on organizations in every sector to measure themselves in relation to comparable organizations both within the country and beyond it. Each organization, indeed each person, faced the new question: what am I doing to contribute to economic competitiveness, especially via innovation in new technologies?

This change included institutions of higher education. Now measurable in terms of national economic competitiveness, institutions that had long produced scholar-officials and other elite workers became subject to methods of quantitative assessment and, ultimately, ranking in relation to other educational institutions, both inside Korea and beyond.

---

[27]   Kunda, *Engineering Culture*, 1992.

[28]   Shin and Kim, "Jabon Habnihwa Undong-Ui Shin Gyeonghyang (The New Trend of Business Rationalization)," 1993, 192; Lee, "Shin Jayuju-Ui Sidae-Ui Munhwa Jabonju-Ui (Cultural Capitalism of the Neoliberal Era)," 2012, 92.

[29]   Koo, "Gonghak Gisulinui Wisang-Gwa Gwaje (Status and Tasks of Korean Engineers)," 1994, 37–39. See also Kim, "Time to Reach a Definite Decision on Technical Workers," 1991.

Figure 5.10: Koo Cha-kyung. Source: LG Group.

A key step was to introduce universities to an image of market competition. In 1995, the so-called "5.31 Education Reform," named for its introduction on May 31, began adjusting funding formulas according to the performance of universities in both education and research. Rather than allocating specific numbers of slots to specific majors at specific universities, the education ministry granted universities new freedom to structure and staff degree programs in order to meet student demand. Students gained permission to enroll in more than one major and even move between universities, transferring their credits. The reform plan also encouraged universities to differentiate their research and teaching functionally. This meant preparing specific cohorts of students for particular categories of future employers and particular sources of research funding.[30]

Engineering curricula across Korea had long emphasized expertise in mathematical problem solving through the engineering sciences. Educational activists saw this attention to more general theory and expertise as inappropriately paying homage to EuroAmerican engineering curricula.[31] As one activist, Lee Ki-jun, a professor at Seoul National University and subsequent president of that university and Minister of Education, put it in 1991, "We have to ... giv[e] up curricula that seemingly relay the development status of the technology-developed countries." Instead, engineering curricula in Korea should "reflect ... demand from domestic industries and create the curricula based on the technology level of Korea. ..."[32] This meant placing increased emphasis

---

[30] Shin, "Gyouk Gaenyeomui Inyeomgwa Cheorak (The Ideas and Philosophy of Educational Reform)," 2005, 10–12.

[31] Lee, "Gonghak Gyoukeun Baljeonhago Inneunga? (Is Korean Engineering Education Developing?)," 1992, 10–12.

[32] Lee, "Gonghak Gyoukui Baljeoneul Wihayeo (For the Development of Engineering Education)," 1991, 356.

on research-based graduate programs and making undergraduate engineering curricula and even technical colleges more "practical." In the new economy, techno-corporate formation would be a contribution to techno-national development.

Most importantly, the large conglomerates began funding new "customized" or "contract-based" university departments focused on providing research and graduates at all levels tailored to the needs of the funding company.[33] In 1994, for example, Daewoo, then the second-largest conglomerate behind Hyundai, funded a new Department of Systems Engineering in the Graduate School of Ajou University. Pohang University of Science and Technology (POSTECH) established the Graduate Institute of Ferrous Technology in concert with POSCO, the steel company. Pusan National University used industry funding to establish the Department of Intelligent Mechanical System Engineering.[34]

In 1996, the Samsung group took a dramatic step deep into the world of higher technical education. It *acquired* a top private institution, Sung Kyun Kwan University (SKKU). The move carried considerable symbolic force because, during the Joseon period, SKKU had produced elite scholar officials, and it still had strong programs in the humanities and social sciences. In one sense, Samsung was simply solving a financial problem. SKKU was in financial distress.[35] In another sense, Samsung was forcefully taking the lead in repositioning university environments to better fit its plans for creative technological development.

Samsung began injecting $100 million into the school each year. It introduced a rating system for faculty productivity in areas ranging from total research funding and publication in quality journals to the attraction and support of quality students. It introduced management principles from the corporate world designed to produce continuous improvement. University employees found themselves evaluated according to the Six Sigma system, designed originally by the U.S.-based Motorola corporation for manufacturing operations. This system challenged its users to minimize "defects" in their work by keeping their rate below three-and-a-half parts per million.[36]

The *chaebols* were particularly aggressive at incorporating into their strategic planning departments and programs at junior colleges, to prepare lower-level technical workers. These emphasized technologies in automobile design and manufacturing, high-tech video media, information systems, shipbuilding, and new materials. The courses they established provided students with hands-on

---

[33] Jeong, "Saneob-Che Witak Gyouk Keun Ingi (Commissioned Education by Industries Wins Popularity)," 1996.

[34] Jeon, "Daehak-Gieop Hyeobnyeok Daegagwon Gwajeong Gaeseol (University and Industry Cooperate to Establish New Department in Graduate School)," 1994.

[35] Kang, "Gwagamhan Tujaga Daehak Hwak Baggeodda (Aggressive Investment Changed University)," 2006.

[36] Lee, "Daehak Hyeogsin 6 Sigma (University Reform 6 Sigma)," 2007; Kim, "Samsungsik Gaehyeok Simnyeon (Samsung-Style Reform over Ten Years)," 2010.

training that they could apply on the job immediately after completion. The number of commissioned departments increased from 42 in 1994 to 76 by 1996.[37]

COMPETITIVE SELF-DEVELOPMENT OR AN ORGANIZED PROFESSION? 1990s

Binding universities more closely to powerful *chaebols* dramatically increased student demand for degrees in engineering. In 1990, Korean universities produced roughly 30,000 graduates of engineering programs. By the end of the decade, this number had increased to nearly 60,000, close to the number produced in the U.S.

The 1990s graduates differed substantially from those in the 1970s in two important ways. In the first place, the newly dominant image of competitiveness challenged them as individuals in ways that anti-communism never had. Resisting the threat of communism brought with it a call, even a mandate, to link arms with others. Communism produced an enemy challenging Korea as a singular whole, making engineers, like others, soldiers working on behalf of their country. The 1990s image of competitiveness, by contrast, not only made individuals more visible as individuals. It also highlighted competition among them, both between countries and within them.

"Second place is not remembered," announced a Samsung advertisement in 1994. Its chairman, Lee Kun-hee, followed by asserting that "One genius can feed millions." Samsung was naming the challenges and opportunities that the Kim Young-sam presidency presented to the *chaebols*. The *chaebols* had to act responsibly, but, above all, make sure they rose to leadership positions in international economic competitions. It was not enough to lead Korea in a world dominated by the image of economic competitiveness. Influential observers of their economic performance were now distributed worldwide.

The Samsung advertisement also named the pressure felt by individuals whose work could be linked to the country's economic positioning and status. Young engineers indifferent to the powerful ideological conflict of the past and raised in a comparatively affluent environment, found themselves challenged to advance individual interests and look for ways to enhance individual competencies. It is in this context that the label "new generation" emerged and spread across the country, labeling the great concerns held by Samsung and other large employers about their employees.

We opened this chapter with excerpts from Samsung's 1993 training manual, which explained that graduates in the 1990s increasingly resisted advice and counsel from older generations on the grounds that such advice no longer applied.[38] Table 5.1 documents how the center of gravity for careers in industry, including engineers, had by 1995 shifted toward newer service-type indus-

[37] Jeong, "Saneob-Che Witak Gyouk Keun Ingi (Commissioned Education by Industries Wins Popularity)," 1996.

[38] Ju, "90nyeondae Hangugui Shinsedaewa Sobimunhwa (The Korean New Generation and Consumption Culture in the 1990s)," 1994, 73.

tries (including here electronics). Between 1975 and 1995, the percentage of employees in service industries had increased from 35% to nearly 55%, with the absolute number more than doubling. Meanwhile the percentage in manufacturing (including heavy and chemical industries) had increased from just under 20% to just under 34%, while the percentage in agriculture, forestry, and fisheries had decreased from over 45% to under 12%. Shifting focus from manufacturing to service industries both authorized and challenged individual engineers to redirect attention away from the country as a whole and onto their employers and, ultimately, themselves.

Table 5.1: Number of employees by industry, 1965–1995

	Total (Unit = 1,000)	Agriculture, Forestry and Fisheries	Manufacturing	Service
1965	8,112 (100.0)	4,742 (58.5)	840 (10.4)	2,530 (31.2)
1970	9,617 (100.0)	4,846 (50.4)	1,377 (14.3)	3,395 (35.3)
1975	11,691 (100.0)	5,339 (45.7)	2,235 (19.1)	4,118 (35.2)
1980	13,683 (100.0)	4,654 (34.0)	3,079 (22.5)	5,951 (43.5)
1985	14,970 (100.0)	3,733 (24.9)	3,659 (24.4)	7,578 (50.6)
1990	18,085 (100.0)	3,237 (17.9)	4,990 (27.6)	9,858 (54.5)
1995	20,414 (100.0)	2,403 (11.8)	6,826 (33.4)	11,185 (54.8)

The reorientation of employees from country to companies became a full-fledged focus on self-development in the wake of the Asian financial crisis of 1997. For our purposes, it is important to know that Korean companies became deeply indebted to banks and that price reductions for Korean goods put in default both many companies and the government itself. Nearly 23,000 businesses vanished in one year, the stock market's price index fell dramatically, and foreign exchange reserves nearly disappeared. The Korean government found itself forced to accept a bailout from the International Monetary Fund in exchange for implementing policies that included the layoff of workers and reductions in wages. Unemployment increased threefold over the next year from roughly half a million workers to a million and a half, and the ratio of non-regular workers increased from 43% to 52%.[39] As Table 5.2 reports, a possible indicator of increased pain was a spike in suicide rates, from roughly 13 per hundred thousand in 1996 to more than 18 in 1998.[40]

Table 5.2: Suicide rates in Korea during the 1990s. Source: www.kostat.go.kr (Statistics Korea)	
Year	**Suicide rate per 100,000**
1991	7.3
1992	8.3
1993	9.4
1994	9.5
1995	10.8
1996	12.9
1997	13.1
1998	18.4
1999	15.0
2000	13.6

A main governmental response, now led by President Kim Dae-jung (1925–2009) (Figure 5.11), elected in 1998, coupled a commitment to reducing unemployment with yet another shift in industrial support. Further emphasizing small service industries and individual initiative, the catchword became "venture businesses."[41] The 1997 Act on Special Measures for the Promotion of Venture Businesses, for example, provided support specifically for entrepreneurial venture businesses, i.e., beyond the purview of the *chaebols*. The initiative successfully increased their number from approximately 2,000 in 1998 to more than 10,000 by 2001.

The financial crisis and increased unemployment further cemented the idea that workers were competing with one another. The *chaebols* and research institutes could no longer guarantee long-term security, and venture businesses by definition championed entrepreneurial individualism. Furthermore, the concise, dense image of "new generation" highlighted by Samsung in 1993 no longer captured all the nuances of shifting attention from community responsibility to self-development.

---

[39] Baek, "Gyeonje Wigi Ihu Hanguk Sahoejeongchaegui Byeonhwawa Hyogwa (Changes and Effects of Social Policy after the IMF Economic Crisis in Korea)," 2011, 88.

[40] KOSTAT, *State Statistics*, 2013.

[41] Jang, "Kim Dae-Jung Jeongbu-Ui Bencheogieop Jiwon Jeongchaeke Gwanhan Gochal (Venture Business Policy under the Kim Dae-Jung Government)," 2005, 29–30.

Figure 5.11: President Kim Dae-jung. Source: National Archives of Korea.

The second way 1990s graduates from engineering degree programs differed from their 1970s counterparts was that they tended, increasingly, not even to identify specifically as engineers or with the practice of engineering. As early as 1993, a leading engineering educator pointed out that while science-technology, or *gwa-hak-gi-sul*, had become widely known and understood as a definable collection of activities, such was not true of the engineering sciences, or *gong-hak*.[42] We should not be surprised. Power in the political economy had shifted away from government and into the private sector. *Chaebols* focused first on competing and surviving in global markets rather than contributing to national development. The prestige of heavy and chemical industries had declined in favor of the new entrepreneurial industries. Creative contributions to new technologies drew upon an increasing range of academic fields. Overall, as graduates from engineering degree programs followed their employers in making themselves competitive in global markets, they gradually lost the ability to claim distinctive identities for themselves as representatives of Korea as a whole. Graduates could no longer hold up knowledge of the engineering sciences as emblematic of Korean progress.

It will not surprise researchers of gender in engineering (and other fields or professions) to learn that, during a decade in which engineers were losing access to elite status as technical scholar-officials, the proportion of women students admitted in engineering colleges and universities tripled in size. Not quite 7% in 1990, by 2000 the proportion increased to nearly 18%.[43] The linkage between the declining status of a field and increased availability to women practitioners is frequently called feminization. In this case, it is difficult to disentangle feminization from the role

[42] Kim, "Hangukui Gonghak Gyouk (Korean Engineering Education)," 1994, 49.

[43] Ministry of Education (MOE), *Gyo-yuk Tonggye Yeonbo* (*Statistical Yearbook of Education*), Annual.

that shifting from manufacturing to service-oriented industries likely played as well. During the 1980s, field positions, overnight work, and work over drinks in manufacturing settings were nearly all limited to men engineers. Women engineers typically found themselves limited to office jobs requiring neat appearance and amiable attitudes. Yet even during the increases of the 1990s, those women who studied science or engineering and pursued careers in industry often found themselves denigrated as "men in skirts" or "honorary men."[44] As accusations, these images labeled women engineers as cold or unsympathetic people who inexplicably put off or gave up gender-appropriate roles of wife and mother in favor of contributing to developments in science and technology.

Meanwhile, for women who sought careers in research through the master's and Ph.D. degrees, the natural sciences actually proved more attractive than pathways through engineering. In 2001, fewer than 10% of master's degrees and 6% of Ph.D.'s went to women engineers. By contrast, the natural sciences attracted 36% of women seeking master's degrees and nearly 25% of those pursuing the Ph.D. Perhaps life in natural science research carried fewer vestiges of the old military chains of command than did the research of engineers wholly targeted to advance industry.

The increases in women seeking the B.S. degree in engineering capped at 18%. It was still that number in 2010. The rates of employment by women may offer some helpful insight. Across the labor market as a whole, the employment of women typically produced an M-shaped pattern. It was highest for women in their early 20s, decreased to a low point during their mid-30s, and then rose again to peak in their mid-40s.[45] Many women evidently insisted on fitting motherhood into lives that also included work. In industry sectors that attracted engineers, however, women produced more of an L-shaped pattern. The precipitous drop presumably indicated that childbirth and child care produced career discontinuities leading to their permanent exit.[46] Many qualified women engineers evidently resisted as individuals the permanent, full-time commitment to career development that the image of men in skirts also conveyed. Industries, it was said, were not much interested in women *gwa-hak-gi-sul-ja*.[47]

---

[44] Nah, "Namnyeo Gonghak Daehakyo-Ui Gunsa Munhwawa Yeohagsaeng Simingwon (Military Culture in Co-Ed Colleges)," 2007, 50.

[45] Park, "Hanguk Yeoseongdeurui Cheod Chui-Eob Jinib Toejange Michineun Saengae Sageonui Yeonghyang (Consequences of Life Events on the First Entry of Korean Women into the Labor Market and Their Subsequent Withdrawal)," 2002, 171.

[46] Kang, "Yeoseong Illyeogwa Gi-Eop Gyeongjaengnyeok (Women Labor and Business Competitiveness)," 2002, 10; Sohn, "Gohagnyeok Jeonmunjik Yeoseongui Nodong Gyeongheomgwa Dillema (Labor Experiences and Dilemmas of Highly Educated Professional Women in Korea)," 2005, 71.

[47] Kim, "Jendeo Ishu Beowi-Ui Hwagjang (Categorical Extension of the Gender Issue)," 2006, 349.

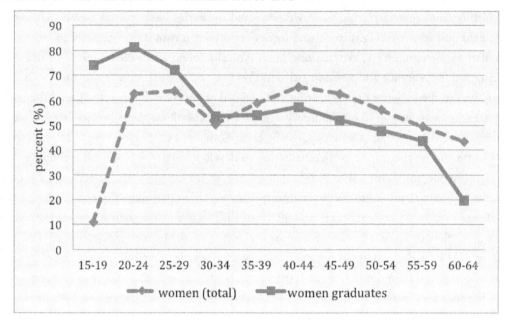

Figure 5.12: Proportions of women working and completing education by age group, 2004. Source: KOSTAT, State Statistics.

Groups of men engineering educators began gathering in the early 1990s to respond to the seeming erasure of pathways to elite status for even their male graduates. Fearing the seeming disappearance of graduate engineers into masses of technical workers in production and research enterprises of various sizes and interests, activist university professors took it upon themselves to formulate and began seeking acceptance for a new image of engineering: it was a "profession." The image of profession explicitly located engineers alongside elite doctors. And perhaps most importantly, it provided a means for engineering educators to promote engineering as (once again!) serving and bearing responsibility for Korea as a whole. A profession worked for everyone.

These activist professors, including the Seoul National University professor Lee Ki-jun mentioned earlier, formed the National Engineering College Deans Council in 1991, the Korean Society for Engineering Education in 1993, the National Academy of Engineering of Korea in 1996, and the Accreditation Board for Engineering Education in Korea in 1998. The Deans Council and Korean Society for Engineering Education formed to increase coordination among deans and faculty, respectively, especially to elevate the status of engineers as leaders within the *chaebols*. The National Academy of Engineering formed to certify the most influential engineering researchers as elite professionals and, in the process, bring education and industry closer together. Its leaders alternate between industry and academia. Two prominent early chairmen from industry included Yun Jong-yong, chairman of Samsung Electronics, and Chung Joon-yang, chairman of POSCO.

Finally, the Accreditation Board was designed to produce greater uniformity in the education of engineers at universities. It gained the support of engineering faculty by promising to improve the chances of engineering rising in status to the level of a profession.

If all went well, these organizations would persuade those conservative engineering faculty who still emphasized general theoretical expertise in the engineering sciences to better target their curricula to the interests of *chaebols*. The first Accreditation Board president reported, for example, hearing frequent complaints from the research institutes and industry about the quality of graduate engineers. "That's why we had to hurry," he asserted hopefully. "[U]niversities . . . do research, foster leaders, and help students establish their value [to employers]."[48] Yoon Woo-young, a member of the Board's committee for external cooperation, elaborated,

> The benefit of [the Accreditation Board] is basically to provide students appropriate for industry. . . . Despite similar admission requirements, graduates of engineering schools become factory managers, while graduates of business schools become CEOs. As professors of engineering schools, this could not be tolerated. There was no single National Assembly member or minister who graduated from engineering school. . . . I definitely thought there was something wrong. It was not the time to say proudly that we properly taught our students their major or mathematics.[49]

Yoon also lamented, "Is it impossible for graduates of engineering schools to become leaders?"[50]

Unfortunately, the answer at the time was, for the most part, yes. Under President Park, graduate engineers had gained access to leadership positions in the national bureaucracy. But that era was sinking further into the past. The populations of graduates who earned degrees in engineering during the 1990s had become too diverse to warrant the designation of a unified profession akin to medicine. So rather than reaching outward beyond arenas of technical work, the activist educators turned their sights inward. They began demarcating graduate engineers from lower-level technicians, the skilled blue-collar workers.

Despite even these efforts, graduate engineers of the 1990s by and large did not look to the professional organizations, including the Accreditation Board for Engineering Education in Korea, for guidance and authority. Lacking the "hungry spirit" that drew them together during the Park era, scaling up an image of a coherent profession was as unlikely as returning to leadership positions in military technocracies. While many older engineers looked back at those days with considerable nostalgia despite the political repression, younger graduate engineers were feverishly competing with one another to find stable employment and ways to lead intellectually and emotionally fulfilling lives.

[48] Interview, Seoul, Korea, March 25, 2013.
[49] Interview, Seoul, Korea, March 20, 2013.
[50] Interview, Seoul, Korea, March 20, 2013.

REFERENCES

Bae, Eong-hwan. "Gwonwiju-Eu Jeongchi Chejeha-Eu Jeongbuwa Gyeongje Iik Jibdan Gwangye (Relationships between Governments and Economic Interest Groups in Authoritarian Political Regimes: The Park Chung-Hee Regime and Chun Doo-Hwan Regime)." *Korean Republic Administation Journal* 35, no. 2 (2001): 19-39. 106

Baek, Doo-joo. "Gyeonje Wigi Ihu Hanguk Sahoejeongchaegui Byeonhwawa Hyogwa (Changes and Effects of Social Policy after the IMF Economic Crisis in Korea)." *Damnon* 21 14, no. 1 (2011): 83-120. 121

Cho, Hwang-hee, Author and Author. *Hanguk-Ui Gwahakgisul Illyeok Jeongchaek* (*Review of Science & Technology Human Resource Policies in Korea*). Seoul: STEPI (Science and Technology Policy Institute), 2002. 109

Cho, Jeong-su. "Saneop-Gye Yogu Insik Suyonggineung Ganghwahaeya (Need to Strengthen the Acceptance of Industrial Demand) " <manuscript>, 1992. 103

Cho, Yu-shik. "Wiro-Buteo-ui Gieop-Hyeok-myeong (Corporate Revolution from the Top)." <manuscript>, 1993. 101

Chung, Chung-kil. "Daetongnyeong-Ui Uisa Gyeol-Jeong-Gwa Jeonmun Gwallyo-Ui Yeokal (Presidential Decision Making and the Role of the Bureaucrat)." *Korean Public Administrative Review* 23, no. 1 (1989): 73-90. 112

Chung, Kun-mo. "Segye Hanminjok Gwahakgisulja Eaehoe Chisa (Speech to Korea Conference on Innovative Science and Technology)." <manuscript>, 1990. 115

Downey, Gary Lee. "The World of Industry-University-Government: Reimagining R&D as America." In *Technoscientific Imaginaries*, edited by Marcus, George, 197-226. Chicago: The University of Chicago Press, 1995. 107

Downey, Gary Lee. *The Machine in Me: An Anthropologist Sits among Computer Engineers*. New York: Routledge, 1998. DOI:10.1525/ae.2000.27.1.182. 107

Hecht, Gabrielle. *The Radiance of France: Nuclear Power and National Identity after World War II.* Cambridge: The MIT Press, 1998. 107

Jang, Ji-ho. "Kim Dae-Jung Jeongbu-Ui Bencheogieop Jiwon Jeongchaeke Gwanhan Gochal (Venture Business Policy under the Kim Dae-Jung Government)." *Korean Public Administratiion Review* 39, no. 3 (2005): 21-41. 121

Jang, Sang-cheol. "Hangugui Gaebal Guk-Ga 1961-1992 (Korean Developmental State: 1961-1992)." Yonsei University, 1999. 106

Jeon, Seong-yong. "Daehak-Gieop Hyeobnyeok Daegagwon Gwajeong Gaeseol (University and Industry Cooperate to Establish New Department in Graduate School)." <manuscript>, 1994. 118

Jeong, Yong-bae. "Saneob-Che Witak Gyouk Keun Ingi (Commissioned Education by Industries Wins Popularity)." <manuscript>, 1996. 118, 119

Ju, Eun-woo. "90nyeondae Hangugui Shinsedaewa Sobimunhwa (The Korean New Generation and Consumption Culture in the 1990s)." *Economy and Society* 21, no. (1994): 70-91. 119

Kang, Hong-jun. "Gwagamhan Tujaga Daehak Hwak Baggeodda (Aggressive Investment Changed University)." <manuscript>, 2006. 118

Kang, Wu-ran, "Yeoseong Illyeogwa Gi-Eop Gyeongjaengnyeok (Women Labor and Business Competitiveness)," Seoul: Samsung Economic Research Institute, 2002. 123

Kim, Chon-wook. "Hangukui Gonghak Gyouk (Korean Engineering Education)." <manuscript>, 1994. 122

Kim, Heung-ki. *Young-Yogui Hanguk Gyeongje (The Glory and Shame of the Korean Economy)*. Seoul: Maeil Business Newspaper & MK Inc., 1999. 112

Kim, Hye-young. "Jendeo Ishu Beowi-Ui Hwagjang (Categorical Extension of the Gender Issue)." Paper presented at the Public Administration Conference, Seoul, 2006. 123

Kim, Il-gyoo. "Samsungsik Gaehyeok Simnyeon, Seongdae Seong Gong Stori (Samsung-Style Reform over Ten Years: The Success Story of Skku)." <manuscript>, 2010. 118

Kim, Kyeong-jun and IMI. "Hyundai Motor Co.: Achieving Independence in Automobile Technology." In *Korea's Seven Top Corporations*, edited by xx-xx. Seoul: One&One Books, 2007. 107

Kim, Lin-su. *Mobang-Eseo Hyeokshin-Euro (Imitation to Innovation)*. Seoul: SIGMA INSIGHT, 2000. 108

Kim, Linsu. *Imitation to Innovation: The Dynamics of Korea's Technological Learning*. Boston: Harvard Business School Press, 1997. DOI: 10.1057/jibs.1997.42. 105

Kim, Seon-hyuk, Ji-ho Jang and Jong-hee Han. "Minjuhwa Ihu-Ui Hangukui Gug-Ga Sahoe Gwangye (State-Society Relations after the Korean Democratization)." Paper presented at the Korean Association for Policy Studies Conference, Seoul, 2006. 115

Kim, Seung-mook. "Urinara Nosa-Gwangye-Ui Sidaebyeol Byeonhwa-E Gwanhan Yeongu (History of Korea Labor-Management Relations from the 1940s to the 2000s)." *Korean Corporation Management Review* 16, no. 4 (2009): 285-309. 110

Kim, Woo-shik. "Korean Engineering Education," Interviewed by Han, Kyong-hee, March 22, 2013.

Kim, Young-woo. "Time to Reach a Definite Decision on Technical Workers." <manuscript>, 1991. 116

Koo, Cha-kyung. "Gonghak Gisulinui Wisang-Gwa Gwaje (Status and Tasks of Korean Engineers)." *Engineering Education* 1, no. 1 (1994): 29-39. 116

KOSTAT. *State Statistics*. <http://kostat.go.kr/portal/korea/index.action%3E, Statistics Korea, Accessed December 20, 2013. 121

Kuk, Min-ho. "80nyeondae Busil Gi-Eop Jeongni Gwajeong-Eul Tonghaebon Guk-Gawa Dae-gi-Eopui Gwangye (The Relationship between the Government and *Chaebols* through the Liquidation Process of Ailing Companies in the 1980s)." *Journal of Modern Social Sciences*, no. 1 (1990): 135-166. 104

Kunda, Gideon. *Engineering Culture: Control and Commitment in a High-Tech Corporation*. Philadelphia: Temple University Press, 1992. 116

Lee, Jae-won. "Daehak Hyeogsin 6 Sigma (University Reform 6 Sigma)." <manuscript>, 2007. 118

Lee, Jang-kyu. *Sillok Yuk Gong Gyeongje (A True Story: The Economy of the Sixth Republic of South Korea)*. Seoul: Joongang M&B, 1995. 111

Lee, Jang-kyu. *Gyeongje-Neun Dangshini Daetongnyeong-Iya (You Are the President in the Economy)*. Seoul: Ollim, 2008. 104

Lee, Ki-jun. "Gonghak Gyoukui Baljeoneul Wihayeo (For the Development of Engineering Education)." *Chemical Industry and Technology* 9, no. 5 (1991): 356-357. 117

Lee, Myun-woo. "Gonghak Gyoukeun Baljeonhago Inneunga? (Is Korean Engineering Education Developing?)." <manuscript>, 1992. 117

Lee, Young-ja. "Shin Jayuju-Ui Sidae-Ui Munhwa Jabonju-Ui (Cultural Capitalism of the Neoliberal Era)." *The Korean Journal of Humanities and the Social Sciences* 36, no. 4 (2012): 87-116. 116

Ministry of Education, Science and Technology, ,, "Gwahaggisul 40 Nyeonsa (Forty-Year History of Korean Science and Technology)," Seoul: Ministry of Education, Science and Technology, 2008. 105

Ministry of Education (MOE), *Gyo-yuk Tonggye Yeonbo (Statistical Yearbook of Education)*, Annual, Seoul (Place Published). 122

Morris-Suzuki, Tessa. *A History of Japanese Economic Thought*. London: Routledge, 1989. DOI:10.1080/08109029008629495. 103

Morris-Suzuki, Tessa. *The Technological Transformation of Japan: From the Seventeenth to the Twenty-First Century.* New York: Cambridge University Press, 1994. DOI:10.1080/0361275 9.1996.9951333. 103

Morris-Suzuki, Tessa. *Re-Inventing Japan: Time, Space, Nation.* Armonk, N.Y.: M.E. Sharpe, 1998. 103

Morris-Suzuki, Tessa and Takur o Seiyama. *Japanese Capitalism since 1945 : Critical Perspectives.* Armonk, N.Y.: M.E. Sharpe, 1989. DOI: 10.2307/2058289. 103

Nah, Yoon-kyeong. "Namnyeo Gonghak Daehakyo-Ui Gunsa Munhwawa Yeohagsaeng Simingwon (Military Culture in Co-Ed Colleges)." *Journal of Korean Women's Studies* 23, no. 1 (2007): 69-102. 123

Park, Byung-young. "Jeongbu Gi-Eop Gwangye-Ui Dayang-Seongkwa Geu Gyeoljeong Yo-in (Government-Business Relations in Korea: Souces of Variation in Textile, Automobile, and Semiconductor Industries in the 1980s)." *Korean Journal of Sociology* 34, no. 4 (2000): 565-595. 103, 106

Park, Jeong-eung. "Sanhak Hyeob-Dong-Gwa Saneop Deahak Gyouk Banghyang (Industry-University Cooperation and Education at Industrial Colleges)." <manuscript>, 1991. 113

Park, Soo-mi. "Hanguk Yeoseongdeurui Cheod Chui-Eob Jinib Toejange Michineun Saengae Sageonui Yeonghyang (Consequences of Life Events on the First Entry of Korean Women into the Labor Market and Their Subsequent Withdrawal)." *Korean Journal of Sociology* 36, no. 2 (2002): 145-174. 123

Shin, Byung-hyun and Do-geun Kim. "Jabon Habnihwa Undong-Ui Shin Gyeonghyang (The New Trend of Business Rationalization)." *Journal of Korean Social Trend and Perspective,* no. (1993): 176-207. 116

Shin, Hyun-seok. "Gyouk Gaenyeomui Inyeomgwa Cheorak (The Ideas and Philosophy of Educational Reform)." *The Journal of Politics of Education* 12, no. 1 (2005): 19-50. 117

Sohn, Seong Young. "Gohagnyeok Jeonmunjik Yeoseongui Nodong Gyeongheomgwa Dillema (Labor Experiences and Dilemmas of Highly-Educated Professional Women in Korea)." *Journal of Korean Women's Studies* 21, no. 3 (2005): 67-97. 123

The Presidential Secretariat. *Kim Young-Sam Daetonglyeong Yeonsiolmunjip 1 (Speech by President Kim Young-Sam).* Seoul: The Presidential Secretariat, 1994. 114

Yoon, Woo-young. "Korean Engineering Education," Interviewed by Han, Kyong-hee, March 27, 2013.

Yun, Jeong-no. *Gwahak Gi-Sul-Gwa Hanguk Sahoe (Science-Technology and Korean Sociey).* Seoul: Moonji Publishing Company, 2000. 105

CHAPTER 6

Engineers for a Post-Catch-Up Korea?

In 2003, the leaders of several advocacy groups for scientist-engineers decided the situation was desperate. They appealed directly to President Roh Moo-hyun for assistance. The advocates represented the Korean Society for Engineering Education, the Korean Federation of Science and Technology Societies, the National Academy of Engineering of Korea, the Korean Council of Deans of Engineering, the Korean Academy of Science and Technology, and the Scientists Association of National Research Institutes.

They visited the Presidential Blue House (*Cheongwadae*) as a group in order to present to the President their "Proposal to Expand Public Office Positions for those with Science and Engineering Backgrounds" (Figure 6.1). Scientist-engineers, according to this group, faced an unprecedented crisis that threatened both their futures and the future of Korea as a whole.

Figure 6.1: President Roh Moo-hyun received representatives of science and engineering sectors in *Cheongwadae*. Source: Presidential Archives.

In this chapter, we examine the emergence of and responses to what scientist-engineers and their advocates came to call the "science and engineering crisis."[1] The crisis was a crisis of identity, techno-national identity. Older scientist-engineers and their advocates continued to campaign for a restoration of elite status as technical scholar-officials. They wanted to be seen and treated as leaders.

Despite some initial successes, their struggle grew increasingly difficult. They failed to disentangle the meaning of technical work, even at the highest levels, from long-standing images that marked work with the hands as low in status. With government encouragement and support, pathways for becoming technical workers and pursuing careers in technical work did continue to multiply. However, consolidating increasingly diverse populations into concisely defined categories that might be judged elite became extremely difficult. Linking these to a compelling vision of a post catch-up Korea became nearly impossible.

SCALING UP AN IMAGE OF CRISIS

Many Koreans in the early 21st century considered November 21, 1997, as a National Humiliation Day. It was the day the country applied for protection and a bailout from the International Monetary Fund (IMF). The name actually called to mind the colonization of Korea by Japan on August 29, 1910. That day had long been called National Humiliation Day. Accepting the IMF Management System in 1997 was humiliating because it signaled inferiority to and dependence on the Western countries that controlled the IMF. Korean strategies to achieve economic parity had not worked. The government's only viable option was to accept external oversight and endure the pains of restructuring the Korean economy.

The so-called "Asian economic crisis" as it was experienced in Korea proved to be especially traumatic for scientist-engineers. In one year, the state-supported workforce in research and development shrunk by nearly 9%. The corporate R&D workforce decreased by nearly 15%. Layoffs and economic stagnation seriously affected professional occupations and labor groups at all levels. Among scientist-engineers, the shock of finding themselves actually expendable affirmed long-standing fears that perhaps life under the Park Chung-hee regime had been an exception rather than a precedent or a beginning. In 1998, the unemployment rate for holders of master's degrees and Ph.D.'s in engineering and the sciences was more than double the national rate of 15.4% By 2000, it reached four times the national rate.[2] A sense of frustration that advocates of engineers had felt throughout the decade spread widely among scientist-engineers, other working engineers, a range of technical workers, and their supporters.

---

[1]   Han, "A Crisis of Identity," 2010.

[2]   Yonhapnews, "Jayeongye Seogbagsa Sireomnyul Simgak (Severe Unemployment Rate among M.S. and Ph.D. Holders in Science and Engineering)," 2004.

In the aftermath of the IMF economic crisis, desires for careers in research and development in both government and industry fell precipitously. The 1990s had witnessed significant expansion. Between the mid-1970s and mid-1990s, the number of researchers with engineering backgrounds had increased more than tenfold, from roughly 5,000 to nearly 80,000. Furthermore, the proportion of graduate engineers working in research rather than manufacturing and other arenas had doubled from roughly one-sixth to one-third.[3]

During the 1990s, however, the supply of graduate engineers to research positions had actually begun to exceed demand. By 1997, neither jobs nor job security were guaranteed to engineering graduates. National humiliation at the hands of the IMF confirmed and emphasized that holders of engineering degrees could no longer expect safe, secure careers. The number of applicants for university degrees in engineering and the sciences (the data are aggregated) fell by 43% between 1998, from 375,000 to under 200,000.[4]

Shockingly, prospective students who had scored high enough on the national exam to gain admission to engineering and the sciences at Seoul National University began to decline offers of admission. In 2002, more than 18% of offers that went out to prospective students came back declined.[5] The best students were moving to medicine, which offered both high status and economic security. Between 1998 and 2001, the percentage of high-scoring students who entered engineering programs decreased from 44% to 33%, while those accepting offers to medical schools increased from 35% to 43%.[6]

In the wake of the financial crisis, the Ministry of Education had introduced a new policy permitting university students to switch freely from one program to another without penalty. Its purpose had been to help them respond proactively to changing job markets and find work. Some students began using the system to apply to programs in engineering and the sciences and then switch to medicine.[7]

An unwelcome spotlight shone on Seoul National University (SNU). A newspaper article published in May 2003 reported: "In order to enter medical colleges that guarantee higher salaries,

---

[3]   Ryoo, "Urinara Gonghak Gisulja-Ui Nodong Sijang (Labor Market for Korean Engineers)," 1997, 228.

[4]   Ministry of Education (MOE), "Gyo-Yuk Tonggye Yeonbo (Statistical Yearbook of Education)," Annual.

[5]   The college entrance in Korea is in general determined by combining two things: scholastic achievement in high school and the scores obtained in the national one-day college entrance examination. People can infer the ranks of universities and disciplines from the combining. If a student applicant receives a few offer letters of admissions from colleges, s/he has to choose one of them, and remaining places will be offered to next applicants. The next applicant has to make the same kind of choice and, therefore, "less popular" places will continue to bid for other applicants. As a result, the rate of registration comes to reflect the popularity of majors and universities among applicants.

[6]   Jin and Yoon, "Godeung Hagsaengui Igonggye Gipi Hyeonsang Siltae Bunseok Mit Gaeseonbangan (The Avoidance of Science and Engineering Majors in Colleges and Universities by Korean High School Students)," 2002, 54–71.

[7]   Kim, "Igonggye Daehak Jinhak Munjejeom Mit Dae-Eung Bangan Yeongu (Problems in University Admissions in Science and Engineering and Their Response Strategies)," 2002,

more and more students are dropping out in the middle of their studies." At SNU, the March to February academic year had just begun, and "a total of 90 students in engineering [had already] submitted withdrawal applications … ." A faculty member affirmed, "When they move to another university, we can generally assume they've gone to medical colleges."[8] That same year, an engineering Ph.D. student at the university publicly shared his fears and regrets about his chosen path: "I will be in my mid-thirties when I get my Ph.D., but research positions have a short life span and it's difficult to change jobs, so they are not all that attractive. These days, I sometimes regret that I didn't apply to medical school."[9] Stories such as this one attracted attention not only across the country but also among media outlets in the West.

Anger blended with shame. Faculty who had themselves been educated during the "golden age" of the 1960s and 1970s found themselves watching the dissolution of engineering into an increasingly heterogeneous arena of education and work. Between 2001 and 2003, public expression of outrage and anguish expanded to include scientist-engineers working at research centers and corporations, engineering and science students at universities, students' parents, and even retired scientists and engineers. In 2003, the president of Chungbuk University summarized the complaint, sense of abandonment, and continuing decline in status relating the stories of an engineer and a graduate in business administration who had returned from prestigious institutions abroad to take up positions at the prominent *Daedeok* Research Park. They started with roughly the same salaries. A decade later, however, the business graduate was earning twice the salary of the engineer and "living in a good neighborhood," while the engineer was "worrying about his children's education."[10]

Like many others, the Chungbuk president glorified the scientist-engineers and lower-level technical workers of the past who "went all over Korea and to the four corners of the world in their grease-stained uniforms and earned a profit for Korea." These technical workers, he elaborated, "spent sleepless nights beyond number at research centers with inferior facilities and gave birth to a technological nation." With "[t]heir sweat … soaked into everything that we enjoy today in such abundance regarding what we wear and eat," he simply could not understand why fewer and fewer students wanted to follow these noble, techno-national footsteps.[11] Given our above analysis of repeated struggles to distinguish industrial work from images that marked work with the hands as low in status, perhaps by calling to mind images of sweat and grease-stained uniforms, the

---

[8]   Yoon and Chun, "Seoul-Dae Igonggye Choe-Agui Jatoe Satae (The Tragedy of Science and Engineering Majors Leaving SNU)," 2003.

[9]   Yoon and Chun, "Seoul-Dae Igonggye Choe-Agui Jatoe Satae (The Tragedy of Science and Engineering Majors Leaving SNU)," 2003.

[10] Shim, "Igonggye Jeolmeunideuri Kkumeul Gat-Gehara (Let Our Young People Have Their Dreams)," 2003.

[11] Shim, "Igonggye Jeolmeunideuri Kkumeul Gat-Gehara (Let Our Young People Have Their Dreams)," 2003.

Chungbuk president may actually have been undermining for many audiences his own claim that such footsteps should be seen as noble.[12]

UNEVEN SUPPORT FROM SUCCESSIVE GOVERNMENTS

Prior to the 2003 trip to the Blue House, advocates for engineers and scientists worked to build justification for official action. On "Science Day" (*Gwahagui Nal*) in April 2002, the Korea Federation of Science and Technology societies formally declared a "science and engineering crisis." It began a petition drive to gain the signatures of a million engineers and scientists in order to seek official redress. A series of articles explaining the crisis appeared later that year in the *Korea Economic Daily*. The "Strong Korea" series won the Samsung Journalism Award, Korea Science Culture Award, and Korea Journalism Award.[13]

President Roh Moo-hyun (1946–2009; in office 2003–2008) responded favorably to the pleas for action embodied in the multitude of informal cries for help and the formal proposal delivered in person. The government framed its actions by linking them to the image of national economic competitiveness that now carried international legitimacy. An initial law to support engineers and scientists was called the "Special Science and Engineering Law for Strengthening the Competitiveness of National Science and Technology" (*Gugga Gwahak Gisul Gyeong Jaeng Nyeok Ganghwareul Wihan Igonggye Teugbyeolbeop*). A second formally elevated the status of scientist-engineers to their highest level in Korean history.[14] Within the administration, a new deputy prime minister for science and technology and a Headquarters for Innovation in Science and Technology appeared to mediate and unify positions across the government on issues involving scientist-engineers and other working engineers.

Most importantly, the government combined separate civil service exams for technology and public administration into the single Higher Civil Service Examination. Successfully passing the new exam would, in principle, open the way for scientist-engineers to compete for a broad array of positions as scholar-officials. It could give them opportunities to identify with Korea as a whole.

---

[12] We must acknowledge here that graduates in the humanities and social sciences beyond economics, political science, and business administration were also struggling mightily to build careers. They no longer had access to the statuses of scholar-officials. In many cases, their challenges were more difficult that those of scientist-engineers. Lee, "Igonggye Wigi Damnone Daehae Dasi Saeng-Gag-Handa (Revisiting the Discourse of Science and Engineering Crisis)," 2002, 77.

[13] Yi, "Gwahak Munhwasang Susangja Seonjeong (Korea Science Culture Award Winner)," 2003.

[14] The law was formally titled the "Supervision System for the Ministry of Science and Technology" (*Gwahak-Gisul Buchongni Jedo*).

Figure 6.2: Kim Woo-sik, Deputy Prime Minister, speaks at NAEK award ceremony. Source: National Academy of Engineering of Korea.

These initiatives quieted the complaints. A researcher at the Korean Institute for Energy Research found it "encouraging" that the government would now "comprehensively implement ..." and coordinate science and technology policies.[15] A KAIST professor optimistically asserted that Korea could become "one of the world's technology powers."[16] Unfortunately, however, the focused official support did not last.

Once the icons of governmental initiatives to define and direct Korea through its heavy and chemical industries, the scientist-engineers (*gwa-hak-gi-sul-ja*) in particular and engineers (*gi-sul-ja*) and lower-status technical workers in general had by the early 2000s become but one collection of special interests and one source of vision for administrative leaders. The Roh government's efforts to restore prominence to engineers and technical workers, while welcomed by those groups, fell short in the midst of economic slump that continued well into the new century. With government no longer in direct control of economic action, supportive policies were not guaranteed realization.

Although the next president, Lee Myung-bak (b. 1941; in office 2008–2013), had been CEO of Hyundai Engineering and Construction, his administration focused on efficiency to end the slump. To the consternation of engineering advocacy groups, it combined the Ministry of Science

---

15 Kim, "Gwagi Buchongnie Baranda (Hope for the Prime Minister for Science and Technology)," 2004.

16 Choi, "Shin Seong-Jang Dongnyeogeul Chajara (Find out the New Growth Engine)," 2004.

and Technology with the Ministry of Education to produce the Ministry of Education, Science, and Technology. This was a demotion. The move deprived the advocates of a powerful unit focused on advancing engineering and science research in favor of one dedicated to greater efficiencies in coordinating education and technology development across the country.

The National Academy of Engineering of Korea, the Korean Federation of Science and Technology Societies, and the Korean Academy of Science and Technology protested. These non-governmental advocates for scientist-engineers attempted to appropriate the image of efficiency. They claimed that efficient economic growth depended upon a "solid infrastructure" of government-industry institutions and high-quality scientist-engineers. To them, the bureaucratic merger was a move "against the stream," and they were "shocked and concerned."[17] Their combined voice had lost status, however. It now had to compete with a chorus of voices approaching education and high technology from a variety of directions.

Five years later, the eldest child of Park Chung-hee, Park Geun-hye (b. 1952) became not only the first woman president but also the first woman leader across northeast Asia. Like her father, Park emphasized unifying the Korean people through national security and a creative economy led by government planning. Yet to formalize planning in the new world of information and computer technologies (ICT), her administration split up the Ministry of Education, Science and Technology and channeled its focus and energy for economic competitiveness through a new Ministry of Science, ICT, and Future Planning.[18]

With education once again gaining its own ministry, engineers, technical workers, and their organizational advocates also found reason to hope yet again that the glory days of the past might return. They would not. The nearly three-decade shift of economic authority away from the government and into the hands of the large conglomerates, the *chaebols*, had altered the meaning and possibilities for centralized planning, let alone elevating technical work to iconic status. Despite the efforts of passionate, hard-working advocates, engineers and technical workers, especially at lower levels, had become increasingly heterogeneous. Dispersed widely across private and public sector jobs, they still confronted versions of the image that work with or close to work with the hands did not warrant consistently elite status. The memory of routine access to high-status positions as scholar-officials had become but a dream.

By 2002, a decade earlier, the powerful ministries, which coordinated governmental initiatives, depended primarily on university graduates in law, economics, and political science. Technology administrators accounted for only a quarter of 88,000 employees. And perhaps more to the point, only 17% of technical administrators held positions in the top three of nine grades. Just

---

[17] Park, "Gwagibu Peji-Neun Choe-Agui Seontaek (Abolition of MOST, the Worst Choice)," 2008.

[18] Park Geun-hye had played the role of first lady after her mother passed away during her father's incumbency. Her policy initiatives suggested she shared many of her father's economic and political aspirations for Korea. Yoon and Kim, "Park Geun-Hye Dangseoninui Lideoshibe Gwanhan Yeongu (The Comparative Leadership Potential of the Korean President-Elect, Park Geun-Hye)," 2013, 81.

as had become the case within the *chaebols*, technical workers in government had become mostly technical support staff.[19]

As we saw in Chapters 3 and 4, which followed the Park Chung-hee era, a generation of manufacturing engineers, scientist-engineers, and other technical workers emerged with the justification to claim they were helping Korea industrialize rapidly and, hence, aspire to take its place among the economically powerful countries of the world. As the lead agents of catch-up growth shifted toward electronic and communications technologies, the engineers and technical workers who emerged found themselves caught in continual struggles for visibility and recognition. By the early 21st century, the champions of the Park era were either retired or nearing retirement. Who, if anyone, would take their place? Would women engineers emerge into leadership positions? How about early-career men engineers? Would they find ways to re-ignite the hungry spirit?

THE CONTINUING STRUGGLES OF WOMEN ENGINEERS

During the first decade of the 21st century, the decreased status of engineering did not have the effect of continuing to attract more women to engineering or of expanding career opportunities for women engineers. The proportion of women graduates from engineering degree programs remained stable at 18% during that period.[20] The ratios of women holding master's degrees remained below 10%, and slightly more than 5% for those holding Ph.D.'s. This compared to roughly 35% and 25% in the sciences.[21]

In research laboratories in both public and private settings, women functioned primarily as "auxiliary human resources."[22] During this period, nearly 65% of women laboratory workers held bachelor's degrees, while slightly more than a quarter had completed master's degrees and only a few percent held a Ph.D. More than 70% of those with master's degrees in either engineering or the sciences worked in universities or government research laboratories. They typically worked as assistants to men professors or laboratory directors holding Ph.D.'s.

Yet these positions were still often preferable to life in private research institutes where they were mostly ranked at lower positions. Perhaps the most striking indicator of their auxiliary, i.e., subordinate, temporary, and replaceable, status in the private sector was that nearly two-thirds of

---

[19] Shin, "Hangugui Gisul Gongmuwon Hyeonhwang-Hwa Gaeseon Bangan (Contemporary Technology Administrators in Korea and Their Development Strategy)," 2003, 8. Government employees in technology are usually assigned to ministries related to certain technologies or industries, and rarely assigned to the Ministry of Strategy and Finance, the Ministry of Planning and Budget, the Ministry of Government Administration, or the Ministry of Education. Shin, "Hangugui Gisul Gongmuwon Hyeonhwang-Hwa Gaeseon Bangan (Contemporary Technology Administrators in Korea and Their Development Strategy)," 2003, 10.

[20] Ministry of Education (MOE), "Gyo-Yuk Tonggye Yeonbo (Statistical Yearbook of Education)," Annual.

[21] Ministry of Education (MOE), "Gyo-Yuk Tonggye Yeonbo (Statistical Yearbook of Education)," Annual.

[22] Lee, "Gwahak Gisulgwa Yeoseong-Ui Jeongchaek Jaeng Jeom (Policies at Issue for Science-Technology and Women)," 2001, 7–8.

women lab workers were in their thirties and almost one-third were in their twenties. Encountering expectations from all directions to leave and fulfill responsibilities of marriage, motherhood, and family, few could even begin to approach the core of R&D activities.[23]

By 2010, the ratio of women among full-time professors in engineering stood at under 4%, noticeably lower than the nearly 21% in the natural sciences.[24] In the government-supported research institutes (GSRIs), the total proportion of women was nearly 12% but only 4% served in senior positions.[25] That by 2011 between 7 and 8% of women workers in the private sector were achieving executive-level positions further illustrates the exceptional struggles that women *gwa-hak-gi-sul-ja* faced in achieving leadership positions in private research work.[26]

This is not to say that women activists were not trying, both inside and outside of government, and in the context of other significant changes across the country. In 2002, the Ministry of Education combined separate seventh-grade courses in "technology" and "homemaking" into one course carrying both names, and required both boys and girls to complete it.[27] When it came into power in 1998, the Kim government established a policy office specifically to encourage "female education." Also, the authorizing legislation for the Ministry of Education in 2000 added a new article titled "Advancement of Equality in Education for Males and Females."[28]

Following the emergence of advocacy organizations for scientist-engineers during the 1990s, the Women in Engineering and Women in Science and Engineering organizations persuaded research universities to better prepare women students for leadership positions (Figure 6.3) and adopt numerical goals to increase the employment of women professors.

[23] Lee, "Gwahak Gisulgwa Yeoseong-Ui Jeongchaek Jaeng Jeom (Policies at Issue for Science-Technology and Women)," 2001, 7.

[24] See below for an argument that links with military training may have distinguished engineering research environments from research work in other fields, including the natural sciences. In addition, the degree in natural sciences includes such departments as food and nutrition science, clothing and textiles science, and home management science, in which 90% of the students are women. Also, students with a natural sciences degree can find teaching positions at *Hagwon* (private educational institutes), middle schools, and high schools. These positions attract more women than men.

[25] Han et al., "Gonghaggwa Jendeo, Gonghak Gyo-Uge Eoddeoke Jeogyonghal Geosinga? (How to Deal with Gender in Engineering Education?)," 2010.

[26] Park, "Yeoseong Imwon Seungjin Haneurui Byeol Ddagi (It Is Too Difficult for Women to Gain Promotions to Executive Status)," 2011.

[27] Chung, "Jeugdeug Gajeong-Gwa Gyoyugui Seong-Gyeoge Gwanhan Seong-Injijeok Jeop-Geun (A Gender-Sensitive Approach to Home Economics in Secondary Education)," 2003, 56.

[28] Kim, "Jendeo Ishu Beomwi-Ui Hwagjang (Extending the Gender Issue into Education Legislation)," 2006, 347.

Figure 6.3: Leadership camp at Yonsei University organized by the local Women in Engineering Program. Source: College of Engineering, Yonsei University.

Policy reports calling for increased participation by women emerged from the Korea Science and Technology Policy Institute, the Korea Development Institute, and the Korea Research Institute for Vocational Education and Training. Finally, clever activists began to popularize the activities and accomplishments of women engineers. In 2011, the chairman of Samsung Electronics drew wide attention by asserting that "women should now become CEOs."[29]

Aggregate statistical indicators suggest that women workers in general achieved little advancement in private companies in the first decade or so after 2000. The proportion of women employees in the 20 largest conglomerates (based on total sales), for example, increased only 2.7% between 2003 and 2013, from 13.9–16.6%.[30] Yet if one begins to parse the data by company, the story of gender composition becomes both multiple and more complex. During this decade-long period, the proportion of women employees at Samsung Electronics, LG Display, and Daewoo

[29] Park, "Yeoseong Imwon Seungjin Haneurui Byeol Ddagi (It Is Too Difficult for Women to Gain Promotions to Executive Status)," 2011.

[30] Kim, "20 Dae Daegieop Yeoseong Biyul Mae-U Naja (Employment of Women in the Twenty Largest Conglomerates Is Very Low)," 2014.

International all decreased, by 4.1, 2.2, and 3.8% respectively. Yet the proportions increased substantially at many companies. At Naver (internet content service provider), SK Networks (textiles, gas stations), and KEPCO (electric power), the proportions increased by 10.2, 16.7, and 12.3%, respectively. Even at Samsung Heavy Industries and POSCO (steel), the proportions increased by 4.3 and 10.6%.[31] Also, anecdotal accounts suggest that in the newest areas of high technology and service industries, women were being employed at greater rates.

The fact that some women engineers in particular locations began to break through glass ceilings and into higher levels of management could indicate that new images of gender were beginning to scale up, redefining the suitability of women engineers for leadership positions and perhaps the meaning of technical work itself. Still, we now turn to another attachment that may have provided continued resistance to professionally ambitious women engineers, the longtime association between engineering practice and military discipline. This association raised the possibility that upward mobility for women in technical careers might best occur around fields of engineering practice rather than through them.

MILITARY PRACTICE AND THE DOMINANT IMAGE OF ENGINEERING

For the U.S. and Soviet Union/Russia, the scaling down of the Cold War and scaling up of economic competitiveness and globalization reduced in significance the ideological confrontation between capitalism and communism. Even China had begun shifting to what would become known as market socialism.[32] But these transformations did not stop the war between the Republic of Korea and the People's Republic of Korea, which technically had not ended with the 1953 armistice.

The South Korean military grew quickly to more than three million personnel. It came to face a North Korean military roughly three times that size, effectively the largest in the world. As South Korean presidents, governments, *chaebols*, university researchers, and entrepreneurial businesses dramatically expanded domestic production and exports, the demands of the military and concerns about invasion from the north continued to provide a dominant source of Korean identity. The border with North Korea was 35 miles from Seoul.

We described in Chapter 3 the first decade of Park's rule, how an image of military discipline infused the regime's centralized, rapid, and impatient infrastructural and economic initiatives. Some scholars have argued that engineers everywhere have depended significantly on images of war for developments in engineering expertise, identity, and commitment.[33] In Korea, engineers and other

[31]   Kim, "20 Dae Daegieop Yeoseong Biyul Mae-U Naja (Employment of Women in the Twenty Largest Conglomerates Is Very Low)," 2014.

[32]   Andreas, *Rise of the Red Engineers*, 2009.

[33]   Blue et al., "Engineering and War: Militarism, Ethics, Institutions, Alternatives," 2013.

technical workers routinely confronted in the workplace images of military practice, and the challenges to their identities posed by those images were distinctly masculine in content.

As recently as the early 2000s, all young Korean men served in the military, unless granted a medical waiver. This in itself is not exceptional. Many countries require military service from young men and, sometimes, women. In Korea, a number of researchers have claimed that the unique military tensions of a divided country, or territory divided into two countries at war, increased the force of military practices on the identities of engineers trained to serve Korea as a whole. The force of the military in Korea, wrote one researcher in the 1990s,

> ...cannot be underestimated in that it has provided a drive for the growth of the nation and society, including enhancement of national and ethnic awareness, respect for the authority and order, a spirit of sacrifice and volunteering, a spirit of cooperation, organizational thoughts, etc. In particular, patriotism, a sense of honor, commitment to organizations, and proactive mind shown by the citizens who have experienced the military became the strongest social asset.[34]

The laboratories and industrial sites of engineering work privileged hierarchy and order. In 2010, a male engineering researcher used military imagery to describe the work environment in his research office: "Do what you are ordered to do." "You can do it." "You must meet the deadline, even by working overnight." "Just do what I order you to do, without making any comments on it." "[If you leave the office] bring the outcome by tomorrow morning." "Why can't we make this [deadline], as others can?" This researcher added, painfully, that he had heard similar sentences from his supervisors in graduate school and, before that, in the military itself.[35]

Park Chung-hee had self-consciously framed his encouragement for industrial developments in militaristic terms, for these presumed the unity he was seeking to create. Students at the initial engineering schools had worked from early morning until late at night. They learned that to "just do it" they had to give their all, all the time. For workers building the Gyeong-bu expressway, led by a military captain, an obstructing hill had become an "enemy." And Park explicitly declared that the purpose of higher education was to "develop ... the nation and natural resources." This could be accomplished only if people understood, accepted, and embodied the challenges of "mental nationality."

Even as industrial development spread beyond the heavy and chemical industries in the 1980s and 1990s, and many engineering graduates from the "new generation" found it difficult to embrace an image of "hungry spirit," vestiges of the military hierarchy continued to operate in R&D work to support the country. A 2007 report on the research culture of the engineering sector documented the effects on women researchers in particular of what it called the "military culture":

---

[34] Park, "Gun Munhwawa Sahoebalgeon (Military Culture and Social Development)," 1991, 64.

[35] BizEng, *Gundae Munhwawa Gwahak Gisul* (*Military Culture and Science-Technology*), 2010.

According to interview results, many professors want women graduate students not to get married or pregnant because they prefer to avoid or delay problems. In some cases, they advise female researchers of an "implied agreement" not to get married while in schools or being employed. . . . Military culture is having a considerably negative effect on the lives and research of female researchers, as well as of male researchers. In graduate schools, female students cannot play a critical role in research activities due to the preference to male students by professors.[36]

At this writing, there is a saying in Korea that a man becomes a real man only after serving in the military. The point is that there he develops a strong sense of service. Numerous researchers have maintained that military order, discipline, and identity fit career pathways through engineering schools and into the large companies that increasingly came to represent the country economically. One argued, for example, that universities in general have been "responsible for educating talents for the nation and companies," have "play[ed] a role in educating male students who have served in the Army into the most competent talents, ... sometimes advocating the military culture centered on returned students." Furthermore, "[t]hose men who graduated from universities and used to serve as officers in the Army are most preferred by companies." The end result is that "the military wants universities, universities want companies, and companies want the military," producing a triangular structure involving all three.[37]

Training engineers to faithfully carry out their responsibilities, the atmosphere in engineering schools also faithfully embodied the associated triangular practices of learning, service, and work. Despite the efforts of all the policy and advocacy groups, the continued dominance of military images of patriotism meant that "male students who have served in the military [have] played a pivotal role in transferring females into second-class members" of society in general but engineering in particular.[38] Successful service through engineering education and economic competitiveness did not fit easily with the identities of those who could not build attachments to the Korean military.

It is in this sense, then, that the early 2000s did not dramatically alter patterns in the career pathways of scientist-engineers, other working engineers, and the larger array of technical workers from the 1980s and 1990s. Engineering did become increasingly heterogeneous and, as such, its status continued to decline. The struggles of advocates faced significant odds against them. At the same time, the continuing dominance of military images in engineering practices simultaneously inhibited engineering and technical work at all levels from becoming increasingly the province of women.

[36] Lee et al., "Yeoseong Yeonguwonui Yeongu Danjeol Choesohwa-Reul Bangan Yeongu (Strategies for Women Researchers to Maintain Continuity in Their Research)," 2007, 6–7.

[37] Nah, "Namnyeo Gonghak Daehakyo-Ui Gunsa Munhwawa Yeohagsaeng Simingwon (Military Culture in Co-Ed Colleges)," 2007, 96–97.

[38] Nah, "Namnyeo Gonghak Daehakyo-Ui Gunsa Munhwawa Yeohagsaeng Simingwon (Military Culture in Co-Ed Colleges)," 2007, 96–97.

NEW IMAGES SCALING UP?

In the midst of decreased public status, ambivalent government support, and continuing attachments to the military images and practices, three novel images also began scaling up among engineers during the early 21st century that warrant close attention in coming years. First, embracing the planet-wide images of economic competitiveness and globalization has challenged engineering educators and employers to redefine what counts as "excellent talents." During the peak years of the 1960s and 1970s, men engineers translated the images of hungry spirit and economy first into challenges to solve problems as quickly and efficiently as possible. The technical problems were clear, and given to them through government directives, and later private initiatives.

At this writing, post-catch-up men and women engineers, by contrast, face challenges to define the problems that should constitute their focus, parse their characteristics, and figure out what range of competencies is necessary and appropriate to address them. Engineers participate in broader conversations and debates about how to build and secure a Korean workforce, now definitely aggregated into a singular whole, that possesses and exhibits excellent talents.

In the midst of such debates, early career engineers appear to put a disproportionately high value on balancing work with life, turning away not only from the techno-national commitments of the 1960s and 1970s but also the techno-corporate commitments of the 1980s and 1990s. By the second decade of the 21st century, it became far more difficult to find examples of the attitude that regards committing oneself to employer and country at the expense of individual interests and desires to be a noble thing to do. Leisure became widely popular, indeed recognized increasingly as productive activity itself. By the same token, images of fatigue became indicators of problems that needed to be dealt with not only by individuals but also at the level of organizations.

Finally, many engineers have interpreted the successes of the "Korean miracle" of the 1960s and 1970s, combined with continuing economic expansion during subsequent decades, as justification for extending their responsibilities beyond the territorial boundaries of Korea itself. At a time when income disparities within Korea remain high, governmental bodies, conglomerates, and engineering organizations have dramatically increased assistance for economic development in the poorer countries of the world. Such work sometimes extrapolates from the image of globalization an expectation to see world problems in development terms and, hence, to understand how others elsewhere view and define development.

Among engineers in EuroAmerican countries, this urge to help is frequently linked to an engineering commitment to help the world as a whole.[39] Across Korea, might new-found wealth play a role? Might engineers envisioning Korea in a position of international leadership be appealing to a still-available image that authentic leadership must include benevolence? Or might wholly new images of Korea's place as a country among countries be scaling up, justifying educational and insti-

---

[39] Lucena et al., *Engineering and Sustainable Community Development*, 2010; Downey, "Normative Holism in Engineering Formation," 2012.

tutional frameworks for understanding, respecting, and accepting responsibilities for the economic struggles of others elsewhere? What might it mean for engineers to add techno-global identities to the configurations of challenges they currently face? How will scientist-engineers, other engineers, and engineering educators reframe their expertise, identities, and commitments as Koreans continue more broadly to re-interpret the experiences of the Park era in relation to both the many formative episodes of the past and the evolving techno-national complexities of the present?

REFERENCES

Andreas, Joel. *Rise of the Red Engineers: The Cultural Revolution and the Origins of China's New Class.* Stanford, Calif.: Stanford University Press, 2009. DOI: 10.1017/S0021911813001873. 141

BizEng. *Gundae Munhwawa Gwahak Gisul (Military Culture and Science-Technology).* http://www.scieng.net, Accessed May 6, 2010. 142

Blue, Ethan, Michael Levine and Dean Nieusma. "Engineering and War: Militarism, Ethics, Institutions, Alternatives." *Synthesis Lectures on Engineers, Technology and Society* 7, no. 3 (2013): 1-121. Morgan & Claypool Publishers. DOI: 10.2200/S00548ED-1V01Y201311ETS020. 141

Choi, Su-moon. "Shin Seong-Jang Dongnyeogeul Chajara (Find out the New Growth Engine)." <manuscript>, 2004. 136

Chung, Hae-Sook. "Jeugdeug Gajeong-Gwa Gyoyugui Seong-Gyeoge Gwanhan Seong-Injijeok Jeop-Geun (A Gender-Sensitive Approach to Home Economics in Secondary Education)." *Korean Home Economics Education Assciation* 15, no. 2 (2003): 55-66. 139

Downey, Gary Lee. "Normative Holism in Engineering Formation." In *Engineering, Development and Philosophy: Chinese, American, and European Perspectives,* edited by Christensen, Steen Hyldgaard, et al., xx-xx. Springer, 2012. DOI: 10.1007/978-94-007-5282-5_14. 144

Han, Kyong-hee, Joon-hong Park and Ho-jung Kang. "Gonghaggwa Jendeo, Gonghak Gyo-Uge Eoddeoke Jeogyonghal Geosinga? (How to Deal with Gender in Engineering Education?)." *Korean Journal of Engineering Education* 13, no. 1 (2010): 38-51. 139

Han, Kyonghee. "A Crisis of Identity: The Kwa-Hak-Ki-Sul-Ja (Scientist-Engineer) in Contemporary Korea." *Engineering Studies* 2, no. 2 (2010): 125-147. DOI: 10.1080/19378629.2010.490557. 132

Jin, Mi-sug and Hyounghan Yoon, "Godeung Hagsaengui Igonggye Gipi Hyeonsang Siltae Bunseok Mit Gaeseonbangan (The Avoidance of Science and Engineering Majors in

Colleges and Universities by Korean High School Students)," Seoul: Korea Research Institute for Vocational Education & Training, 2002. 133

Kim, Hye-young. "Jendeo Ishu Beomwi-Ui Hwagjang (Extending the Gender Issue into Education Legislation)." Paper presented at the Public Administration Conference, Seoul, 2006. 139

Kim, Hyo-jin. "20dae Daegieop Yeoseong Biyul Mae-U Naja (Employment of Women in the Twenty Largest Conglomerates Is Very Low)." <manuscript>, 2014. 140, 141

Kim, Ju-hoon, "Igonggye Daehak Jinhak Munjejeom Mit Dae-Eung Bangan Yeongu (Problems in University Admissions in Science and Engineering and Their Response Strategies)," Seoul: Korea Educational Development Institute, 2002. 133

Kim, Yo-sep. "Gwagi Buchongnie Baranda (Hope for the Prime Minister for Science and Technology)." <manuscript>, 2004. 136

Lee, Eun-kyung, "Gwahak Gisulgwa Yeoseong-Ui Jeongchaek Jaeng Jeom (Policies at Issue for Science-Technology and Women)," Seoul: STEPI, 2001. 138, 139

Lee, Su-young, Mi-sug Jin, Seon-mee Shin and Young-min Lee, "Yeoseong Yeonguwonui Yeongu Danjeol Choesohwa-Reul Bangan Yeongu (Strategies for Women Researchers to Maintain Continuity in Their Research)," Seoul: Korea Research Institute for Vocational Education & Training, 2007. 143

Lee, Young-hee. "Igonggye Wigi Damnone Daehae Dasi Saeng-Gag-Handa (Revisiting the Discourse of Science and Engineering Crisis)." *Gyo-yuk Bipyeong* 8, no. 0 (2002): 75-79. 135

Lucena, Juan, Jen Schneider and Jon A. Leydens. *Engineering and Sustainable Community Development.* San Rafael, California: Morgan & Claypool Publishers, 2010. DOI: 10.2200/S00247ED1V01Y201001ETS011. 144

Ministry of Education (MOE), "Gyo-Yuk Tonggye Yeonbo (Statistical Yearbook of Education)," Seoul: Annual. 133, 138

Nah, Yoon-kyeong. "Namnyeo Gonghak Daehakyo-Ui Gunsa Munhwawa Yeohagsaeng Simingwon (Military Culture in Co-Ed Colleges)." *Journal of Korean Women's Studies* 23, no. 1 (2007): 69-102. 143

Park, Bang-ju. "Gwagibu Peji-Neun Choe-Agui Seontaek (Abolition of MOST, the Worst Choice)." <manuscript>, 2008. 137

Park, Cheol-geun. "Yeoseong Imwon Seungjin Haneurui Byeol Ddagi (It Is Too Difficult for Women to Gain Promotions to Executive Status)." <manuscript>, 2011. 139, 140

Park, Jae-ha, "Gun Munhwawa Sahoebalgeon (Military Culture and Social Development)," Seoul: Korea Institute for Defense Analyses, 1991. 142

Ryoo, Jae-woo. "Urinara Gonghak Gisulja-Ui Nodong Sijang (Labor Market for Korean Engineers)." *Korean Journal of Labor Economics* 20, no. 2 (1997): 221-254. 133

Shim, Bang-wung. "Igonggye Jeolmeunideuri Kkumeul Gat-Gehara (Let Our Young People Have Their Dreams) " <manuscript>, 2003. 134

Shin, Mun-ju. "Hangugui Gisul Gongmuwon Hyeonhwang-Hwa Gaeseon Bangan (Contemporary Technology Administrators in Korea and Their Development Strategy)" *Engineering Education* 10, no. 4 (2003): 7-12. 138

Yi, Chae-rin. "Gwahak Munhwasang Susangja Seonjeong (Korea Science Culture Award Winner)." <manuscript>, 2003. 135

Yonhapnews. "Jayeongye Seogbagsa Sireomnyul Simgak (Severe Unemployment Rate among MS and PhD Holders in Science and Engineering)." <manuscript>, 2004. 132

Yoon, He-shin and In-sung Chun. "Seoul-Dae Igonggye Choe-Agui Jatoe Satae (The Tragedy of Science and Engineering Majors Leaving SNU)." <manuscript>, 2003. 134

Yoon, jong-sung and young-oh Kim. "Park Geun-Hye Dangseoninui Lideoshibe Gwanhan Yeongu (The Comparative Leadership Potential of the Korean President-Elect, Park Geun-Hye)." *Social Science Research Review* 29, no. 1 (2013): 71-93. 137

CHAPTER 7

Engineers and Korea

"My competitive advantage against other women colleagues" lies in in the extent to which "I [am] an employee who … diligently work[s] to improve the productivity of the company, not as a housewife or mother." So said in 2014 an engineer who had achieved the status of managing director for a major electric power company. Having earned her Ph.D. in mechanical engineering from Seoul National University, she elaborated that she had spent "most of my available time 'improving my performance' so I can be recognized as an expert." She had to do her best, she said, because "my company fairly pays me for what I do for it."

The managing director characterized her greatest strength as "external activity," or building positive networks with many people. She reported that men typically learned such practices during their military service, especially through parties and "drinking occasions." Women, therefore, had to avoid being seen as "individualistic," meaning not reaching out to others. Indeed, because this engineer considered other people's positions so much, she had earned the nickname "Excessive Consideration."[1]

"I like the process of detecting and solving problems," said another engineer, a graduate in mechanical engineering from the private Yonsei University, in early 2014. He was describing his research on safety systems for a car manufacturer to reduce the effects of collisions. He "[felt] rewarded and a sense of responsibility, for I know how my job affects society." He had been inspired in middle school by a teacher who had demonstrated a "candid and sincere attitude" and an "honest way of … life, without any pretending." He was happy to serve his company, which "helps its employees a lot."

Still, he was feeling great pressure "to learn something by myself." He felt challenged in particular "to understand and have an open attitude toward language proficiency and different cultures." This engineer also made a point of asserting that his life extended beyond his work. He "spen[t] all weekends only for myself," and although he was open to working "a little late if necessary," he mostly fulfilled his job responsibilities "during working hours" alone. Indeed, he did not expect his connection to the company to be permanent. If he could further develop his "competency" as an individual engineer, he firmly believed he could "find any job and do well at any company that I want."[2]

These two stories make visible some of the images challenging graduate engineers working in the Republic of Korea during the early 21st century. When a highly capable woman engineer

[1] Interview, Seongnam, Korea, April 9, 2014.

[2] Interview, Seoul, Korea, March 21, 2014

earned a Ph.D. and rose to the level of managing director in her company, she became a kind of scholar-official for the private sector. One can see in this trajectory her responses to both long-established images of authority and a newer image of economic competitiveness. One can also see responses to economic competitiveness and, perhaps, even to an image of globalization in the other engineer's plans to develop competencies in language and culture, this time so he can move on to another job with another company.

Dominant images of gender still seemed operative. The managing director had to overcome expectations she would be too "individualistic" and thus fail to integrate well with men colleagues who had learned how to network in the military. Competing with women colleagues who still comprised with her a minority, she had actually over-emphasized connectedness, earning for herself a new nickname. Also, the image of advancing the country through the private sector showed up in the work of the automotive engineer who proudly demonstrated his commitment to "society" via his responsibility to the company.

In Chapter 1, we committed to introduce you to what it has meant to be an engineer across the Republic of Korea. We promised to address five organizing questions, including: (1) How and why did Korea create new categories of technical workers, including engineers? (2) How did engineering education and engineering practices emerge? (3) Who has gained the opportunity to claim the identity of engineer and who has not? (4) How have engineers been educated and trained and where have they tended to work? (5) What emerged as key issues for engineers and engineering across South Korea during the early 21st century?

We made a number of choices in providing what must be considered partial answers to these questions. We made these choices to address different types of readers at the same time. For researchers, this account of the fraught attempts of engineers to embrace Korea as a whole raises the possibility that localized versions of "techno-national formation," or desires for such, may be regularly, if not always, at stake in the expertise, identities, and commitments of engineers. For engineering students and working engineers, our goal remains encouraging you to move beyond seeking straightforward information about what it has meant to become and be an engineer in Korea. By focusing on what engineers and engineering have been *for* across Korea, we are trying to help you better reflect on and analyze your own expertise, identities, and commitments, whether as Korean or non-Korean engineers, and in interaction with others, including both engineers and non-engineers.

Our main methodological strategy has been to follow the emergence of dominant images of engineers and engineering, making visible changes and continuities at the same time. As we built a historical map, or geography, of emergent engineers and engineering, each of our organizing questions became, in fact, a distinct question at different points in time. To address and answer them, we found it essential to explore the evolving image of the scholar-official and struggles of engineers and their advocates to gain access to it.

KOREAN ENGINEERS AND THE SCHOLAR-OFFICIAL

Not until the years following the 1950s Korean War did a ruling government express explicit interest in producing engineers and other technical workers born in the Korean peninsula. The efforts of Park Chung-hee and his government during the period 1961–1979 stand out as dramatic initiatives to produce (co-produce) simultaneously a powerful Korea that would take its place in the world alongside other powerful countries and a hierarchical array of technical workers who would lead the development of this new Korea. The figure of the engineer gained prominence just at the moment that a transformational movement began to remake the Republic of Korea into a potentially unified nation. As the extensive literature on engineering formation and national identity suggests, this connection between the emergence of the engineer and emergence of the nation is by no means unique.

Note also that the new categories of technical worker that Park's government created gained meaning and significance in relation to previously established expertise and identities. This also was not unique. In this case, long-established class identities stood out in particular as key realities with which would-be makers of engineers would have to deal.

SCHOLAR-OFFICIALS WITH SUPERIOR VIRTUE

The Dasan (1762–1836) episode during the late Joseon dynasty illustrated how members of the *yangban* aristocracy demonstrated their superior virtue and qualification for elite status in ruling bureaucracies. They learned to read aloud and interpret the Four Books and Five Classics of Neo-Confucian thought and practice. Dasan had completed this work as an adolescent, passed the literary exam that gained him admission to the national Confucian academy, and then the higher civil service exam that earned him a position in the Office of Royal Decrees. He had become a scholar-official, clearly superior to farmers, artisans, and merchants who had no such expertise.

But Dasan resisted the attachments that came with this identity. Even prior to British colonial interventions, traders and missionaries from that empire as well as from Portugal, the Netherlands, and France had made it clear to officials of the Qing dynasty that powerful civilizations lay beyond its boundaries. Dasan's choice of resistance through construction projects and advocating technologies for industry and national defense challenged the subordinate statuses not only of the artisan but also of the Joseon dynasty and its territory.

Dasan formulated and attempted to scale up an image of Joseon identity that would build new attachments to the empires of the West. Without challenging dominant images of a righteous government and more moral society, Dasan added to neo-Confucianist expectations a commitment to collective material progress via technical work. Yet he failed to persuade other scholar-officials that Koreans were lazy not to make machines. Other members of the *yangban* aristocracy were surely aware of the European traders and missionaries and their many new machines, but these did not provide justification for scaling down the educational formation that had made them aristocrats in the first place. The Dasan episode affirmed that practices of synthetically manufacturing nature were pedestrian activities subordinate to the spiritual, purifying process of memorizing and attempting to follow the prescriptions of sacred texts.

This is not to say that scholar-officials of all sorts were ignoring opportunities to integrate Western learning, especially medical practices, into their own work. We are not asserting that a shared, grammar-like, deep-seated culture of neo-Confucianism was resisting change. Indeed, probably everyone who encountered European traders and missionaries found themselves changed somehow in the process. Rather we are saying that neither Dasan nor others who may have been like-minded were able to elevate the status of technical learning and practice to an extent that competed, even remotely, with the memorization of sacred texts. For most of its history, the Joseon dynasty did not produce workers who might claim the identities of engineers.

It became much easier to argue for new practices of technical learning after the Japanese first threatened invasion in 1875. The Opium Wars in China of 1839–1842 and 1856–1860 had made clear to governments across northeast Asia that every territory was subject to interventions, unequal treaties, and the humiliation of extraterritoriality (free mobility and trade without duties) imposed by Europeans and Americans. The new Japanese empire was cleverly appropriating Western knowledge and techniques from, especially, the British. It had begun protecting Japanese territory from external intrusion by extending its own reach.

King Gojong tried to follow the Japanese lead, which included producing and privileging engineers to advance its empire. He detached Joseon from China by establishing a Korean empire and initiated efforts to integrate practices of Western industry. His officials sent students to Japan, established industrial schools for commoners and lower-level aristocrats, and created the National Agricultural, Commerce, and Technical School to produce technical workers who might achieve the status of scholar-officials. Meanwhile, missionaries formed private schools to prepare workers for industry. Neither collection of efforts reached the point of producing engineers, however. These initial practices of techno-imperial formation created by the state and of techno-evangelical formation established by Christian churches were sharply reduced when the empire became a colony.

During the Japanese colonial period, the Governor-General and supporting officials sought to eliminate the Korean scholar-official completely by dissolving it into a massive, homogeneous class of people—inferior Koreans. As the Japanese government worked to eliminate all vestiges of Korean autonomy, difference, and identity, it limited technical education to elementary courses, and technical work to insecure, temporary positions on colonial projects. Even the Kyeong-sung Higher Engineering School trained low-level technicians and managers to support Japanese-led industry.

Lacking a domestic image to which they could aspire, Korean students of means who sought the prestige of higher education typically traveled to Japan. They usually gained access only to technical secondary schools. Christian missionaries in the province updated their techno-evangelism through private colleges, but these could not offer higher status or reliable employment. The Governor-General did establish an imperial university to provide higher-skilled labor in medicine, law, and administration. Yet the purpose was to export the new knowledge and techniques of Japanese medical and bureaucratic leaders. A significant majority of its students were Japanese.

The number of students seeking mid-level technical education within the Joseon province did quadruple between 1919 and 1935, from roughly 3,000 to more than 12,000. Yet the proportion of Koreans in that population never exceeded 25%. The Korean scholar-official remained sharply suppressed as the Japanese empire extended its large industrial conglomerates, the *zaibatsu*, into the province, and also used it to host and support a Continental Military Logistics Base.

After the Pacific War, the occupying U.S. military unwittingly supported a reactionary vision of a rural country led by a landlord class and scholar-officials. Forcing together ten different educational institutions in order to transfer and scale up an American image of efficient higher education, U.S. officials sent a strong message that they expected the Republic of Korea to be highly centralized. Yet nationalist movements during the war, coupled with readily available socialist images of a new world led by workers, had fueled widespread dreams of a people's Korea. A flood of technical workers went north to help achieve liberation for subordinated masses through large-scale industrial development led by a technocratic Communist Party. A new national assembly in South Korea did begin promoting both industry and technical education in 1948. Funding was simply unavailable, however, to produce technical practitioners who could seek the status of scholar-officials. Then the North invaded.

The scholar-official did lose its rural attachments when land reform, necessary to keep up with techno-socialism in the North, capped ownership at 7.5 acres. The *yangban* elite adapted by embracing new attachments to business. University attendance shot up from 8,000 in 1945 to 100,000 in 1960, but the national assembly kept technical education subordinate to academic education in the humanities and social sciences. Only the latter could foster appropriate moral and spiritual development. During the 1950s, the U.S government promoted technical education, including engineering, through the Minnesota Project and support for atomic energy, but these targeted efforts did not achieve a broad reach. The small, light industries that technical entrepreneurs developed did not prompt nor justify a national commitment to higher education in engineering.

SCALING UP TECHNO-NATIONAL FORMATION

As the position of scholar-official gained and lost identities from the Joseon dynasty to the late 1950s, its detachment from work with the hands remained intact. The persistent low status of technical work, even that associated with new industrial practices, provided an important backdrop to Park Chung-hee's vision for a new Korea. Clean, smooth hands, Park famously argued, were those of the privileged class that had to be replaced via a reconstruction of Korea. Giving visibility and prestige for the first time to technical workers, from semi-skilled technical labor to research scientist-engineers, Park's program sought to transform technical education into genuine techno-national formation.

Technical workers gained elevated status as contributors to a national workforce, made official in 1962 in the first of several classification systems. The neologism *gisul* enabled industrial work to have high status by distinguishing it from the hands-on labor of artisans. The Professional Engineers Act of 1963 repositioned engineers (*gi-sul-ja*) into the realm of scholar-officials with the new category *gi-sul-sa*, to label male scholars with authority. The powerful Economics Planning Board and, later, Ministry of Science and Technology led a panoply of state actions to promote domestic industry. During the 1960s, however, the main driver of economic expansion was through small, light industries, including those that depended upon the low-cost labor of women whose technical training was limited to primary education.

The heyday for engineers across the Republic of Korea came during the 1970s. The second phase of President Park's program shifted emphasis to the heavy and chemical industries, and the Ministry of Science and Technology championed a new category of scholar engineer, the *gwa-hak-gi-sul-ja*, or scientist-engineer. These moves gave engineers the identities of technical literati who performed and led creative activities. After graduating from a four-year science and engineering college or university, they would take positions in charge of research and development, planning, and management, and, ultimately, develop new, specifically Korean forms of science-technology (*gwa-hak-gi-sul*). The focus was never on basic science. Many of these new technical literati actually did become scholar-officials, using expertise in calculation to administer laws and policies to promote industrial development and expand exports.

By the mid-1970s, with government actively supporting family-owned industrial conglomerates in a variety of heavy and chemical industries, scientist-engineers technicians, and craftspersons could actually project a novel present into a transformational future. Those with higher education could imagine positions of official leadership, and those at lower levels could imagine significant upward mobility—if not for themselves, then for their children.

However, Park's new images never scaled up sufficiently to become taken for granted. He was working against other, powerful new images of opportunity and status that had nothing to do with techno-national formation and exports. Adding new content to neo-Confucian education, students and families of means continued to place high value on moral and ethical training, now including comprehensive grounding in history, literacy, math, and general science. Enrollments in technical schools did increase during the 1960s, but enrollments in academic tracks increased far more. By the end of the 1970s, the Park government had successfully attracted many young men and women from rural areas into the cities, challenging them with the hungry spirit and calling them the flag-bearers of national modernization. But Park never overcame resistance to his vision of national reconstruction, and those who claimed the identity of engineer would find their attachments to Korea as a whole to be a fleeting embrace.

NEW VISIONS OF TECHNO-NATIONAL FORMATION

President Chun faced dramatically new circumstances when energy prices increased rapidly, and the country experienced negative economic growth for the first time. His government argued that the Park preference for picking companies had been inefficient, even irrational. To rationalize the economy, the Chun government would shift the locus of imagination and control from the state to the private sector, help minimize overlaps in production among *chaebols*, and rectify an overreliance on heavy and chemical industries. Yet rather than withdrawing support for industry, the government broadened and diffused it by shifting funds away from specific industries and into infrastructural support and basic technologies, to stimulate a wide range of industries.

The shift of initiative and industrial leadership to the private sector had the cumulative effect of strengthening the *chaebols*. Many were already increasing their R&D activities because U.S. and European companies increasingly viewed East Asian conglomerates as formidable economic competitors. That a shift was taking place became obvious as early as 1980 when Hyundai refused to leave the automobile business.

Shifting the locus of imagination and control enabled new visions of techno-national formation to garner greater support. The first to achieve prominence was techno-democratic formation. In the face of growing resistance to political repression, government leaders sought to embrace an image of governance in which their principal role changed from directing to supporting, especially small- and medium-sized enterprises in the rapidly-expanding computer and electronics industries. Increasing attendance by students in engineering became one part of a huge increase in university enrollments across the country. Taken as a whole, the main label for these 1980s moves was "democracy." Withdrawing from direct control of the economy would help produce a country and modes of educational formation that were more democratic.

A second, more lasting image to scale up centered on techno-corporate formation. Within the large conglomerates, neither graduate engineers nor other technically trained employees had grounds for claiming special status other than technical experts. The identities, allegiances, and commitments of engineers shifted from embracing the country as a whole to hopefully benefitting the country by embracing the commitments of employers. The managing director and safety engineer introduced above provide cases in point.

President Kim Young-sam's civilian government accelerated the growth of this vision. His "new economy" called for individual creativity within both large and small organizations, in both private and public sectors. Broad infrastructural support would not enable Korea to catch up to powerful Western countries. In its place, the Kim approach introduced coordinated action between universities and *chaebols*.

Under Chun, engineering faculties had focused on teaching, and they retained control over curricular contents. This changed during the 1990s. Students gained new freedoms to choose their curricula, universities began adding research institutes, engineering became more accessible to women students, and higher education faced new metrics that linked them to the economic competitiveness of the country as a whole. In the process, *chaebols* gained significant influence over engineering education and research. Moves to fund customized and contract-based departments peaked in 1996 when Samsung acquired and then actively managed Sung Kyun Kwan University.

The techno-corporate approach to techno-national formation did direct engineers to align their personal commitments with those of their employers. But an equally significant change was that it challenged them to compete more directly with one another as individuals, in the same way that their country competed with other countries.

Perhaps not surprisingly, an increasingly common response by early-career engineers was to turn away from identifying themselves with the country as a whole, and sometimes even from the commitments of their employers. Many came to see themselves as working, first and foremost, for themselves and their families, making competencies and career calculations as much a matter of varying personal or family commitments as the fulfillment of broader responsibilities to employer, country, and even humanity writ large. Massive layoffs during the Asian financial crisis contributed to this inward turn.

Increasing detachment from broader commitments and greater emphasis on self-development produced increased heterogeneity in the expertise, identities, and commitments of graduate engineers. Many stopped identifying as engineers, and, in the early 2000s, many of the highest-qualified candidates for engineering programs turned to medicine, law, or economics.

Activist engineering educators had formed new organizations during the 1990s to scale up yet another approach to techno-national formation—preparation for a profession. But graduate engineers of the 1990s and early 21st century by and large did not look to these organizations for guidance in building careers and fulfilling responsibilities to their employers and the country as a whole. Furthermore, participation by women in engineering did not continue to grow, possibly because of longstanding tight linkages between engineering and military practices and identities. At this writing, some early career engineers, raised in comparative affluence, have begun extrapolating an embrace of Korea to include others elsewhere by adding techno-global identities to both educational formation and trajectories of work.

In sum, the three decades following the peak of uniform techno-national formation under President Park Chung-hee had the cumulative effect of separating most graduate engineers from access to the status and identity of the scholar-official. This counted as a loss not only for the most elite scientist-engineers (who, like the managing director above, might find rough equivalents in the private sector) but also for the lower-level technicians and other technical workers who found it increasingly difficult to see themselves as more than updated versions of hands-on artisans. The energetic efforts of advocates promoting professionalism for engineers continued unabated. But separation from the figure of the scholar-official brought increased disappearance into the depths of huge conglomerates, a shift of technological focus to electronics and communications industries over which many new fields claimed jurisdiction,[3] and ever-expanding heterogeneity in the work identities of graduate engineers.[4] Forced to respond to a dominant image of economic competitiveness and, later, globalization, the techno-national formation of Korean engineers lost its uniformity and gained multiple purposes or aims.

CRITICAL SELF-REFLECTION AND CRITICAL PARTICIPATION

There is no single culture of Korean engineers. That is, there is no single language-like grammar located deep in the bellies of engineers educated or formed across the Republic of Korea. It is essential to keep this in mind as one wrestles with and implements the term "engineering cultures." What is an engineering culture if not an organized collection of deep-seated assumptions that structure what those do who become engineers within the Republic of Korea?

We have drawn here upon a different concept of culture that, we maintain, enabled us to better attend to both the dramatic differences among people who have claimed the identity of engineer across Korea and the difficult, evolving challenges they have shared. In this approach, an engineering culture consists of historically emergent images of engineers and engineering, nominated by advocates. As specific images of engineering expertise, identities, and commitments scale up to dominance, engineering students and working engineers respond in ways that are idiosyncratic and historically contingent yet also frequently patterned. The outcome in this account is what we call the techno-national formation of engineers.

Our account has many holes. We have not attempted to map in detail patterns of curricula and curricular formation for *gwa-hak-gi-sul-ja* (scientist-engineer) or *gi-sul-ja* (engineer), let alone *gi-neung-ja* (technician) or *gi-neung-gong* (craftsman). The account does not call attention to the hierarchies that live within these categories, or the multitude of ways in which actual career pathways blur the distinctions. We have mentioned only Seoul National University, Yonsei University, and Sung Kyun Kwan University, leaving untouched and unexamined the multitude of different ways that higher education in engineering has been positioned and carried out. We did not dive into local institutional histories or call attention to regional differences. While we have traced some flows of meaning and practice from China, Japan, the U.S., and Europe, we have not followed transfers of images, practices, people, and things in significant detail. Much remains to be explored.

What we have done is to provide one map of the emergence of engineers and engineering across a terrain whose main identity changed numerous times from the late Joseon Dynasty to the contemporary Republic of Korea. We have produced a historical geography of techno-national formation across Korea.

---

[3] Abbott, *The System of Professions: An Essay on the Division of Expert Labor*, 1988.

[4] Williams, *Retooling*, 2002.

For readers who are engineering students or working engineers, you are building personal geographies by encountering career challenges and making career choices. Having read this book, we ask you to consider how the historical geography that it contains relates to the personal geography you are building. How would you define and measure your expertise, identity, and commitments in relation to the evolving approaches to techno-national formation reported in this book? What might techno-national, techno-imperial, techno-corporate, techno-democratic, techno-evangelical, or techno-global expertise, identities, and commitments mean to you? Can you learn about yourself by juxtaposing the pathways engineers have followed in this account with the pathway you are following yourself? Can reflecting on the emergence of engineers and engineering across Korea help you to better reflect critically on the choices and judgments you have made in responding to challenges born both within engineering curricula and across the territory that you count as your own?

Furthermore, if you find yourself in contact with engineering students or working engineers educated in Korea, are you now able to ask more informed and sophisticated questions about how they are located, what they know, and what they want? Indeed, might reading this book increase your level of interest in and curiosity about the expertise, identities, and commitments of any others with whom you work, including both engineers and non-engineers? To the extent others educated and trained in ways different from you carry with them different approaches to defining what counts as significant problems, might you as an engineer be able to use this book to improve your own ability to define problems effectively with others, again including both engineers and non-engineers? Collaborative problem definition that recognizes differences in power, expertise, and design is, we maintain, essential to quality global engineering.

Finally, for those of you who are scholars seeking critical participation in the making of engineers, whether in Korea or elsewhere, a take-away message of this book is that localized efforts to intervene in the formation of engineers should at least question the extent to which the engineers that concern you define their expertise, identities, and commitments in techno-national terms. In the event the answer is in some sense yes, then one must calculate to the best of one's ability just what sorts of meanings one's identities and initiatives carry for engineers and how these might fit or conflict with what existing techno-national commitments they deem important. When confronted with new images of their education and work, engineers will measure them against what they already embrace and take for granted. Any outcomes from the changes you seek to accomplish will likely be a mix of changes and continuities in dominant images of who engineers are and what they do. The trick is to figure out and persuade others exactly what this mix could be.

REFERENCES

Abbott, Andrew. *The System of Professions: An Essay on the Division of Expert Labor*. Chicago and London: The University of Chicago Press, 1988. 149

Williams, Rosalind H. *Retooling: A Historian Confronts Technological Change.* Cambridge, Mass.: MIT Press, 2002. 149

Index

Accreditation Board for Engineering Education in Korea (ABEEK), 124, 125

Advancement of Equality in Education for Males and Females Article, 139

Agrarian country, 38, 41, 48

Anti-communism, 38, 39, 41, 54, 57, 70, 101, 116, 119

Appledore Shipbuilders, 79, 80

Asian Games, 110

Atlantic Baron, launching ceremony, 81

Automobile Industry Long-term Promotion Plan, 86

Banners, 19, 77

    engineer is bearer of the nation's industrialization, 77

    banner at Busan Mechanical Engineering High School, 78

Battelle Memorial Institute, 63

Battelle, Gordon, 63, 64

Biological Research Information Center, 3

Catch-up growth, 115, 138

Catholicism, 23

    and egalitarian society, 24

CDMA

    see Code Division Multiple Access

Ceremony for Delivering Relief Supplies, 45, 46

Chaebol, 83, 87, 116, 124, 137

    concept of, 82

    educational influence, 118, 125, 156

    employment, 90, 108, 112, 113, 119, 121

    governmental policy, 103, 106, 115

    international competitions, 119, 122

    relations with government, 83, 101, 104, 115, 155

Chang, Kai-shek, 38

Choi, Hyung-sub, 64, 85, 92

Christian Colleges, 34-35

    Yonhi College, 34

    Soongsil Academy, 34

Ewha School for Women, 35

Chun Doo-hwan, President, 102, 108-112

Chung, Joon-yang, 124

Chung, Ju-yung, 79, 80, 104

Classification system of technical workers, 8-12

 classification by the Economic Planning Board, 8-9, 57, 59-61, 65, 68, 89

 classification by the Ministry of Science and Technology and 1973 Qualification Act, 11

 classification of research workers by the Ministry of Science and Technology, 12

 craftsman (*gi-neung-gong, gi-neung-ja*), 8-12, 58, 88, 157

 engineer (*gi-sul-ja, gong-hag-in*), 8-9, 11-13, 58, 65, 88, 136, 154, 157

 field technician (*hyeon-jang-gi-sul-ja*), 10, 11

 industrial engineer (*san-eob-gi-sa*), 11, 88

 learner (*gyeon-seub-gong*), 9

 master craftsman (*gi-neung-jang*), 11, 88, 90

 professional engineers (*gi-sul-sa*), 11, 58, 65, 88, 154

 research assistant (*yeongu josu*), 10, 11

 scientist (*gwa-hak-ja*), 11, 12

 scientist-engineer (*gwa-hak-gi-sul-ja*), 10, 11, 13, 65

 semi-skilled craftsman (*ban-sug-lyeon-gong*), 9

 semi-skilled production worker, 8

 skilled craftsman (*sug-lyeon-gong*), 9

 technician (*gi-sul-gong*), 8, 9, 58, 66

 technologists of industrial machinery, 8

Code Division Multiple Access, 105

Cold War, 42, 107, 141

Confucianism, 17, 24

 dominant ideals of the *yangban* literati, 30

 Four Books and Five Classics, 17, 151

 image of nature, 26

 images of leadership, 86, 116

 neo-Confucian images of superiority through literary purification, 30, 36

 purify one's way, 25

Continent Military Logistics Base, 35, 153

Corporate Culture White Paper, 115

Corporate culture, 115

Critical participation, 158

Critical self-reflection, 157

Customized university departments issues, 118, 156

Daedeok Research Park, 134

Daewoo, 80, 104, 105, 118, 140

Dakaski Research Institute, 40

Dasan

    see Jeong, Yak-yong

Declaration of Crisis in Science and Technology, 1

Defense industry, 78, 87

Democratization, 19, 101, 112

    democratization movement, 109

Deputy Prime for Science and Technology, 135, 136

Dominant images

    concept of, 5

    dominance of military images, 143

    dominant commitments, 5

    dominant practices, 5, 102

    dominant image of a disaggregated Korea, 116

    dominant image of anti-communism, 116

    dominant image of economic competitiveness, 107, 119, 157

    dominant images of engineers and engineering, 7, 150

    dominant image of filial piety, 23

    dominant images of gender, 7, 150

    dominant image of the humanist scholar-official, 93, 112

Dong-do-seo-gi, concept of, 31

Doosan Infracore, 2

East Asian financial crisis, 102, 120, 121, 133, 156

Economic competitiveness, 3, 19, 116, 137, 143

    image of competitiveness, 7, 101, 107, 119, 144, 150, 157

    national economic competitiveness, 116, 135

Economic Planning Board, 8, 57, 59, 60, 61, 65, 89

"Economy First", 54, 57, 64, 66, 67, 110, 144

Education, academic, 42, 61, 71, 153

    fever for, 72

    growth during the 1950s, 42-43

    Joseon dynasty, 31

Education reform, 115

       5.31 Education Reform, 117

Electronic Telecommunications Research Institutes (ETRI), 105, 106

Electronics Industry Promotion Act, 86

Employment Training Centers Law, 88

Engineers, Korean

       generational differences, 2-3, 101

       concepts, 7-13

       loss of commitment, 2

       scholarly engineers, 58

       sense of defeat, 3

       sense of frustration 1-3

Engineering, Korean

       concept, 9

       profession, 119, 124, 125, 156

       reverse engineering, 84-85

Engineering education, Korean

       curricular issues, 113, 117-119

       engineering science (Gong-hak), 122

Engineering Cultures, 115

Engineering workforce

       definition, 13

Export-led growth, 55, 66, 102

Family-owned businesses, 82, 154

Federation of Korean Industries, 114

Five-year plan for economic development, 59

Gender in engineering, 122-123

       gender composition of graduates, 12

       honorary men, 123

       men in skirts, 123

gi-sul (technology), 29-30

Globalization, 141, 157

       image of globalization, 144, 150

Gorbachev, Mikhail, 107

Government-supported Research Institutes, 139

Ground-breaking Ceremony for the Ulsan Industrial Area, 59, 60

Growth of university graduates issues, 90

GSRIs
      see Government-supported Research Institutes
Gwa-hak-gi-sul-ja
      see Scientist-engineer
Gyeong-bu Expressway, 70, 93
      burning incense, 71
Hangeul, 17, 67
Hanyang University, 2
Headquarters for Innovation in Science and Technology, 86
Heavy and chemical industries, 86-89, 93, 120, 122, 136, 154
Higher civil service examination, 135, 151
Hong, *Dae-yong*, 26
      Ji-jeon-seol, 26
"Hungry Spirit" (hungry Jeong-shin), concept and issues, 1-2, 125, 138, 142, 144, 154
Hyundai Engineering and Construction, 136
Hyundai Group, 79
Hyundai Heavy Industry Company, 80
Hyundai Motor Company, 80, 104
Industrial Development Act, 105
Industrial Education Promotion Act, 61, 88
Industrial Generic Technology Development Project (IGTDP), 105
Industrial Soldiers, 62
International Bank for Reconstruction and Development, 70
International Competition, 107
International Monetary Fund (IMF), 132, 133
Invention Society, 36
Jang, Yeong-sil, 96
ja (human), 8
Japan
      archipelago, 17, 27, 28
      colonization, 35, 68, 132
      dominant image of strength and virtue, 27
      Ordinance on Joseon Education, 31
      policies on Joseon education, 31-33
      recognition of western civilization, 27-28
      samurai, 26, 27
      shogunate, 27, 28

Tokugawa household, 26, 27

Jeong, Yak-yong (Dasan), 23-24, 26, 151-152

    pontoon bridge, 23

    fortress, 24

    pulley, 24

Johnson, President Lyndon 63

Joseon dynasty

    education, 29

    geopolitics, 23

    social classes of (*Sa-nong-gong-sang*), 17

Joseon Daily Editorials, 35

Keiretsu, 82

KEPCO (electric power), 141

Kim Ilsung University, 40

Kim, Chang-eop, 25

Kim Dae-jung, President, 121, 122

Kim, Jae-ik, 112

Kim, Woo-sik, 136

Kim, Yong-kwan, 36

Kim Young-sam, President, 112, 114, 116, 119, 155

    civilian government, 113, 115, 116

    mental featism, 114

    "New Korea", issues, 114-116

Kim, Yun-shik, 30

King Gojong, 26, 152

King Sejong, 96

KIST

    see Korea Institute of Science and Technology

Koo, Ja-kyung, 116, 117

Korea

    Democratic People's Republic of Korea (North Korea), 14

    Great Korean Empire (Dae-han-je-guk), 17, 26, 28, 31, 47

    identity of, 37, 47, 79, 141

    liberation from Japanese empire, 38

    peninsula, 5, 10, 23, 26, 29, 48, 54

    Republic of Korea (South Korea), 14, 17, 48, 141

    Sovereignty issues, 14-18

three kingdoms: Silla, Baekje, and Goguryeo, 15-16

    unified, 57, 86, 114, 151

Korea Development Institute, 140

Korea Institute of Science and Technology (KIST), 63, 64, 86, 92

Korea Military Academy, 111

Korea Research Institute for Vocational Education and Training, 140

Korea Science and Technology Policy Institute, 140

Korea Telecommunications, 105

Korean Academy of Science and Technology, 131, 137

Korean Central Intelligence Agency, 101

Korean Federation of Science and Technology Societies (KOFST), 1, 131. 137

Korean miracle, 93, 94, 144

Korean Office of Atomic Energy, 43

Korean Society for Engineering Education, 124, 131

Korean War, 8, 24, 44, 151

KSEE

    see Korean Society for Engineering Education

Kwon, Oh-kyung, 2, 3

Kyeong-seong Higher Engineering School, 33, 35, 37

Kyeong-seong Textile, 36

Kyeongseong Imperial University, 39

Kyoto Imperial University, 40

Labor Unions, 110

    labor disputes, 110

    labor strike at Ulsan District, 110

Land reform, 67, 72, 153

Lee, Hyun-soon, 2

Lee, Ki-jun, 117, 124

Lee Myung-bak, President, 136

Lee, Soon-shin, General, 80

Lee, Young-woo, 87

LG, 105, 107, 116, 117

    LG display, 140

Li, Seung-ki, 40, 41

Light industries, 54, 62, 77, 83, 94, 153, 154

Lucky Goldstar

    see LG

Machinery Industry Promotion Act, 86

Mao, Zedong, 38

March 1 Movement, 33

Meiji Restoration, 27, 30, 67, 78

Military, Korean

       culture, 142, 143

       discipline, 78, 86, 114, 141

       hierarchy, 19, 142

       practice, 141-143

       rule, 55, 113

       triangular structure, 86

Ministry of Education, 39, 70, 72, 89, 95, 133, 139

Ministry of Education, Science, and Technology, 137

Ministry of Science and Technology, 10, 11, 12, 65, 66, 86, 96

Ministry of Science, ICT, and Future Planning, 137

Minnesota Project, 43, 153

Mitsubishi, 107

Modernization, 2, 69, 70, 94

Mongolian empire, 16

Monopoly Regulation and Fair Trade Act, 106

Movement to Promote Science to the Whole Nation, 96

National Academy of Engineering of Korea, 1, 124, 131, 137

NAEK

       see National Academy of Engineering of Korea

National Agricultural, Commerce, and Technical School, 28, 31, 32, 152

National Charter of Education, 68, 69, 72

       Declaration of the Charter, 69

National Engineering College Deans Council, 124

National Humiliation Day, 132, 133

National identity

       collective, 95, 97

       concept of, 6, 151

       connections between engineers and, 6

       self-consciousness, 66

       technological developments and, 6

National R&D Project, 105

National Science and Technology Commission, 115

National Science Foundation, 40, 65

National Technical Qualification Act, 88

National Technical Training Center, 32

Naver (Internet Content Service Provider), 141

Neo-Confucianism

      see Confucianism

Neoliberalism, 103, 107

"New economy", 114, 115, 116, 155

"New Generation", 101, 102, 119, 121

New Village Movement, 96

Nixon, President Richard, 78

Nishe, Amane, 30, 65

O, Won-chul, 89, 90, 91, 92

Opium wars, 152

Pacific War, 35, 37, 38, 42, 48, 72, 153

Park Chung-hee, President, 17, 53-54, 56, 63, 77, 113, 132, 151, 156

      development dictator, 78, 79

      economic strategy, 102

      export-oriented policy, 62

      General, 53, 55, 57

      mental nationality, 95, 101, 142

      Our Nation's Path: Ideology of Social Reconstruction, 66

      repression of critics and political opponents, 93

      restoration (*Yushin*), 78

      strategic goal and numerical goals, 59, 60, 93

      The Country, The Revolution, and I, 53

      triangular configuration of military-style leadership, 86

      vision, 153

Park Geun-hye, President, 137

Park, Tae-jun, 86, 87

Pohang Iron and Steel Company, 83, 86

      Ignition Ceremony at Second Blast Furnace, 84

Pohang University of Science and Technology, 118

POSCO

      see Pohang Iron and Steel Company

POSTECH

      see Pohang University of Science and Technology

Post catch-up, 132, 144

Presidents, Korean
 Chun Doo-hwan (1980-1988), 102
 Kim Dae-jung (1998-2003), 122
 Kim Young-sam (1993-1998), 114
 Lee Myung-bak (2008-2013), 136
 Park Chung-hee (1963-1979), 56
 Park Geun-hye (2013-2018), 137
 Rhee Syng-man (1948-1960), 82
 Roh Moo-hyun (2003-2008), 131
 Roh Tae-woo (1988-1993), 112

Qing dynasty (China), 17, 32

Rhee Syng-man, President, 45, 46, 54, 57, 66, 82, 91
 resignation, 59

Roh Moo-hyun, President, 131
 government's efforts, 135

Roh Tae-woo, President, 111, 112
 ordinary people, 67, 111

President Ronald Reagan, 103
 Reaganomics, 103

Presidential Advisory Committee on Education Reform, 115

Presidential Blue House, 68, 131

Presidential Council on Intellectual Property, 2

Private-public Collaborations, 79
 public and private constituencies, 91

Private-sector initiatives, 103
 private-led market competitive and open-style economic operations, 106

Professional Engineers Act, 58, 154

Proposal to Expand Public Office Positions for those with Science and Engineering
 Backgrounds, 131

Railroad line, 28

Rationalization of the economy, 102-103

Ree, Tai-kyue, 39, 40 ,41

Sa, Gong-il, 112

Samsung, 118, 121, 135, 156
 1993 training manual, 119
 Samsung Electronics, 124, 140

Scholar-official, 10, 11, 17, 29, 58, 61, 65, 72, 88, 101, 109, 112, 116, 118, 135, 137, 150, 152, 154, 156

      elite status, 23, 47, 93

      intellectual resources, 25-26, 30, 31, 152

      landed gentry, and (yangban), 23, 153

      scholar administrators, 61, 112

      technical scholar-officials, 91, 122, 132, 154

Science and Engineering Crisis

      issues, 132-134, 135

      national discussion and debate, 1

Science and Technology Annals, 9, 11, 12

Science and Technology Promotion Law, 65

Science Day, 1, 37, 135

Scieng.net Forum, 3

Scientist-Engineer (*gwa-hak-gi-sul-ja*), 3, 53, 73, 79, 96, 102, 112, 131, 139, 143, 156

      concept of, 10-13, 65, 66, 154

      crisis discourse, 132-135

      image of, 90, 95, 101, 136, 137, 138

Scientists Association of National Research Institutes, 131

"Second Economy," 18, 54, 69, 70

      economy of spirit, 96

Seoul Olympic Games, 111

Seoul National University, 39, 40, 41, 43, 48, 68, 117, 124

Service Industries, 141

Seth, Michael, 72

      Fever for academic education, 72

Shipbuilding Industry Promotion Act, 86

Shipbuilding Industry, 80

Sibal, 47

Sidehara, Taira, 32

Six Sigma System, 118

SK Networks, 141

Social Darwinism, 36

Soviet Union, 38, 40, 43, 48, 107, 141

      Stalinist, 54, 59

Special Committee for National Security Measures, 103

Special Measures for the Promotion of Venture Businesses Act, 121

Special Science and Engineering Law for Strengthening the Competitiveness of National Science and Technology, 135

Spring of Seoul, 113

Statistical Yearbook of Education, 12

Steel Industry Promotion Act, 86

Suicide Rates, 121

Sung Kyun Kwan University (SKKU), 156, 157

Sunkyoung (SK Group), 85

Support of Specific Research Institutes Act, 85

Supreme Council for National Reconstruction, 57, 61, 68

Technical experts, 91, 113, 155

Technical leaders, 31, 91

Technical literati, 65, 73, 154

Technical Services Support Act, 85

Technical soldiers, 57

Technicians, low-level, 8-13, 19, 24, 29, 31, 33, 35, 40, 48, 58, 61, 66, 89, 101, 125, 152, 154, 156, 157

Techno-national formation, 8, 10, 33, 40, 71, 88, 154, 155, 156

    concept of, 18, 44, 150, 157

    country building projects, 7

    secondary and higher technical education, 9, 10, 54

    techno-corporate formation, 113, 118, 144, 155, 156

    techno-educational progress, 96

    techno-educational vision, 101

    techno-evangelism, 152

    techno-global identities, 145, 156

    techno-imperial, 152

    techno-national complexities, 145

    techno-national development, 118

    techno-national experts, 91

    techno-national identities, 113, 132

    techno-national icons, 101

    techno-national leader, 77

    techno-national scholar, 11, 65

    techno-national status, 58, 62

    techno-national whole, 67

Technocrats, 91-92, 97

    technical bureaucracy, 95

Technology, Korean concepts of, 8
 li-yong, 25
 gae-mul, 25
 gi-sul, 29-30, 31
Technology administrators, 137
Technology Development Promotion Act, 85, 105
Telegraph, 28
 first telephone exchange, 29
Telephone, 28
Thatcher, Margaret, 103
 Thatcherism, 103
Tram, 28
Turnkey, 83
U.S. Military Government in Korea, 38, 39
Ulsan plant, 104
United Nations, 38, 44, 46
University-based research, 115, 116
University of Minnesota, 43
USAMGIK
 see U.S. Military Government in Korea
Venture businesses, 121
Vietnam War, 63
Vocational education, 43, 61
Vocational Training Regulation Law, 88
Women in engineering, 138-141, 156
 auxiliary human resources, 138
 executive-level positions, 139
 gender composition of graduates, 12, 140
 honorary men, 123
 men in skirts, 123
 organizations, 139
 proportions of women working and completing education by age group, 124
 senior positions, 139
 women engineering students, 123, 138
 women scientist-engineers (women *gwa-hak-gi-sul-ja*), 123, 139
 Yonsei university camp, 140
Women in Science and Engineering Organization, 139

Women workers, 8, 13, 58, 66, 139, 140
      dutiful daughters, 62
      entrance rate to high schools, colleges, and higher technical education, 89
      factory girls, 62
World Bank, 70
Yangban, (rural aristocrats), 23, 30, 31, 32, 65, 72, 82, 87, 151, 153
      concept of, 17
      educational process, 34, 36, 47
      political power, 33, 42, 54
      thoughts, 25-26
Yamaga, Shinji, 37
Yeon-Haeng-Rok Selection, 25
      emissaries, 25
Yonsei Unversity, 149
Yoon, Woo-young, 125
Yuge, Kotaro, 32
Yun, Jong-yong, 2, 124

Author Biographies

Kyonghee Han received her Ph.D. from the Department of Sociology at Yonsei University and is now an assistant professor in the Engineering Education Innovation Center at Yonsei University. She teaches Engineering and Society and Engineering Ethics in the College of Engineering. She has conducted research on how the social roles and identities of engineering and engineers have formed and changed. Her recent research examines how engineers have recognized and changed their sense of social responsibility in relation to a series of technological controversies that have taken place in Korea. She also develops and operates various programs to promote the innovation of engineering curricula.

Gary Downey is Alumni Distinguished Professor of Science and Technology Studies and affiliated professor of Women's and Gender Studies at Virginia Tech. A mechanical engineer (B.S., Lehigh) and cultural anthropologist (Ph.D., University of Chicago), he is the author of *The Machine in Me* and Co-Editor of *Cyborgs and Citadels* and *What Is Global Engineering Education For?* He edits the *Engineering Studies Series* (MIT Press), *Global Engineering Series* (Morgan & Claypool Publishers), and *Engineering Studies* journal (Routledge/Taylor & Francis). He is co-founder of the International Network for Engineering Studies, as well as founder of the Engineering Cultures course. He serves as President of the Society for Social Studies of Science (2013-2015).

Printed in the United States
by Baker & Taylor Publisher Services